AF787286

Nuclear Data

Applications to society and industry

Online at: https://doi.org/10.1088/978-0-7503-5102-7

IOP Series in Nuclear Spectroscopy and Nuclear Structure

Series Editors

John Wood, Georgia Institute of Technology, USA
Kristiaan Heyde, Ghent University, Belgium
David Jenkins, University of York, UK

About the Series:

The IOP Series in Nuclear Spectroscopy and Nuclear Structure provides up-to-date coverage of theoretical and experimental topics in all areas of modern nuclear spectroscopy and structure. Books in the series range from student primers, graduate textbooks, research monographs and practical guides to meet the needs of students and scientists.

All aspects of nuclear spectroscopy and structure research are included, for example:

- Nuclear systematics and data
- Ab-initio models
- Shell model-based descriptions
- Nuclear collective models
- Nuclear symmetries and algebraic descriptions
- Many-body aspects of nuclear structure
- Quantum mechanics for nuclear structure study
- Spectroscopy with gamma-rays, charged particles and neutrons following radioactive decay and reactions
- Spectroscopy of rare isotopes
- Detectors employed for gamma-ray, neutron and charged-particle detection in nuclear spectroscopy and related societal applications.

Nuclear Data

Applications to society and industry

Edited by
David G Jenkins
School of Physics, Engineering and Technology, University of York, York YO10 5DD, UK

IOP Publishing, Bristol, UK

© IOP Publishing Ltd 2025. All rights, including for text and data mining (TDM), artificial intelligence (AI) training, and similar technologies, are reserved.

This book is available under the terms of the IOP-Standard Books License

No part of this publication may be reproduced, stored in a retrieval system, subjected to any form of TDM or used for the training of any AI systems or similar technologies, or transmitted in any form or by any means, electronic, mechanical, photocopying, recording or otherwise, without the prior permission of the publisher, or as expressly permitted by law or under terms agreed with the appropriate rights organization. Certain types of copying may be permitted in accordance with the terms of licences issued by the Copyright Licensing Agency, the Copyright Clearance Centre and other reproduction rights organizations.

Certain images in this publication have been obtained by the authors from the Wikipedia/ Wikimedia website, where they were made available under a Creative Commons licence or stated to be in the public domain. Please see individual figure captions in this publication for details. To the extent that the law allows, IOP Publishing disclaim any liability that any person may suffer as a result of accessing, using or forwarding the images. Any reuse rights should be checked and permission should be sought if necessary from Wikipedia/Wikimedia and/or the copyright owner (as appropriate) before using or forwarding the images.

Permission to make use of IOP Publishing content other than as set out above may be sought at permissions@ioppublishing.org.

David G Jenkins has asserted their right to be identified as the author of this work in accordance with sections 77 and 78 of the Copyright, Designs and Patents Act 1988.

ISBN 978-0-7503-5102-7 (ebook)
ISBN 978-0-7503-5100-3 (print)
ISBN 978-0-7503-5103-4 (myPrint)
ISBN 978-0-7503-5101-0 (mobi)

DOI 10.1088/978-0-7503-5102-7

Version: 20251101

IOP ebooks

British Library Cataloguing-in-Publication Data: A catalogue record for this book is available from the British Library.

Published by IOP Publishing, wholly owned by The Institute of Physics, London

IOP Publishing, No.2 The Distillery, Glassfields, Avon Street, Bristol, BS2 0GR, UK

US Office: IOP Publishing, Inc., 190 North Independence Mall West, Suite 601, Philadelphia, PA 19106, USA

Contents

Preface

Nuclear physics—the discipline that focuses on the properties and interactions of the atomic nucleus—is a mature field where a great deal of understanding has been accumulated over the last century. The applications of nuclear physics to industry and society are wide ranging and impactful in areas as diverse as energy generation and healthcare. Our ability to safely and efficiently exploit such nuclear applications, however, is founded in a detailed knowledge of *nuclear data*. Nuclear data can mean many things in this context from basic properties of a given isotope such as its half-life and decay modes, to the cross section (or energy-dependent probability) for different nuclear reactions. This book sets out to shine a light on this often under-appreciated domain of nuclear data. In doing so, it will focus on a range of different industrial and societal applications, starting with the emerging domain of nuclear fusion before discussing several different aspects where nuclear data are important to nuclear fission energy and the nuclear fuel cycle. Lastly, the book will cover topics related to medical isotope production. The book opens with an introduction to key topics in nuclear structure, nuclear reaction theory and nuclear data evaluation which underpin the other chapters. This material is pitched at a relatively high level but extensive references are provided throughout allowing the reader to drill into particular topics in more detail.

This book would not have been possible without the contribution of colleagues working across both pure and applied nuclear physics. Having such a diversity of inputs allows us to approach the same material from different perspectives. In addition, the majority of the contributors to this book are based in the United Kingdom. This naturally provides a particular focus for the topics presented here. It also reflects the long-standing heritage of nuclear physics and its applications in the UK; key milestones of which include Rutherford's experiments in Manchester which led to the hypothesis of the existence of the nucleus of the atom in 1911, and the opening of the world's first fission reactor for energy generation at Calder Hall in 1956.

The hope is that this book will allow postgraduate and advanced undergraduate students to gain an appreciation of how the nuclear physics they learn is applied to real world applications. In addition, this book will be invaluable to those starting work in any industrial or medical setting where their work will concern nuclear data.

Editor biography

David G Jenkins

David Jenkins is Head of the Nuclear Physics Group at the University of York in the UK. In 2023, he received the Rutherford Medal and Prize from the Institute of Physics for *outstanding contributions to experimental nuclear physics and commitment to widening participation*. His research in experimental nuclear physics focusses on several topics such as nuclear astrophysics, clustering in nuclei and the study of proton-rich nuclei. In recent years, he has developed a strong strand of applications-related research with extensive industrial collaboration. In particular, he was worked in nuclear data studies and developed bespoke instrumentation for nuclear decommissioning, homeland security and medical imaging.

List of contributors

Alejandro Algora
University of Valencia, Valencia, Spain

Greg Bailey
UKAEA, Culham, UK

James Benstead
AWE, Aldermaston, UK

Mamad Eslami
University of York, York, UK

William Gelletly
University of Surrey, Guildford, UK

Mark Gilbert
UKAEA, Culham, UK

David Jenkins
University of York, York, UK

Giuseppe Lorusso
National Physical Laboratory, Teddington, UK

Lee Packer
UKAEA, Culham, UK

Andrew Petts
EDF, Hartlepool, UK

Jagjit Singh
The University of Manchester, Manchester, UK

Alex Valentine
UKAEA, Culham, UK

Chapter 1

Nuclear theory

James Benstead, David Jenkins, Jagjit Singh and William Gelletly

This introductory chapter presents some of the key nuclear theory which will assist in our understanding, extraction and manipulation of nuclear data. The relevant theory falls broadly under two headings—nuclear structure and nuclear reaction theory—although as will be seen, there is a strong interplay between the two. For example, the details of nuclear structure can be followed through to their implication for the probability of different nuclear reactions occurring. This chapter begins with nuclear structure theory, initially focussing on gross nuclear properties such as mass and binding energy before turning to nuclear excited states and nuclear decay modes. This leads on to a basic introduction to nuclear reaction theory focussing on some of the key approaches. Finally, we will look at how nuclear data are extracted and compiled.

Concepts: Gross properties: mass, radii, nuclear binding energy; excited states; decay modes: α, β, γ, fission; nuclear reactions; cross sections; compound versus direct reactions; wavefunctions; optical model, DWBA model, *R*-matrix, Hauser–Feshbach model, nuclear data evaluation

1.1 Nuclear structure

Nuclear structure is a major topic in its own right and there is only scope here to address some of the aspects of most relevance to the goals of the present volume. Those who have already covered a reasonable amount of nuclear physics in undergraduate courses may find the upcoming material somewhat basic and want to skip ahead. On the other hand, the reader who seeks a more comprehensive introduction to nuclear structure is directed to earlier volumes in this e-book series, in particular *Nuclear Data: A Primer* [1], and might consider tackling the more advanced volumes which focus, in more detail, on the underlying structure of excited states in nuclei: *Nuclear Data: A Collective Motion View* [2] and *Nuclear Data: An Independent-Particle Motion View* [3]. All three *Nuclear Data* volumes are data-driven; they do not start with nuclear models as many textbooks do, but rather with

doi:10.1088/978-0-7503-5102-7ch1

UK Ministry of Defence
© Crown owned copyright 2025/AWE.

the extensive databases related to nuclear structure such as the Evaluated Nuclear Structure Data File (ENSDF) maintained by the National Nuclear Data Center (NNDC) in the US. In a sense, these earlier volumes parallel what the present volume sets out to achieve in terms of nuclear data but with a focus on pure science rather than societal applications.

1.1.1 Gross properties

Let us begin by exploring what might be termed the *gross properties* of nuclei, i.e. their mass and radius, since such observables give us a high level picture as regards the interplay of the forces between nucleons (i.e. protons and neutrons) within the nucleus. An overview of the principal forces and interactions is presented in figure 1.1. The most significant interaction is the strong nuclear force which is responsible for forming nuclei in the first place since it attracts nucleons together at short range. The strong force is also repulsive at very short range preventing the nucleus from simply imploding. In principle, the nuclear force is expected to exhibit higher-order effects, i.e. its strength between individual nucleons is affected by the presence or absence of additional nearby nucleons. Nuclei are finite in size and so there are also corrections to strong-force interactions at the nuclear surface, where there are fewer surrounding nucleons to feel the attraction of the strong interaction. Protons and neutrons differ in that protons have a charge while neutrons are uncharged. Coulomb repulsion between protons in the nucleus will, to some extent, counteract the strong interaction. Such an effect might be expected to become more important as the number of protons in the system increases; indeed, the increasing instability of high-Z nuclei is what limits the known periodic table to 118 elements with the heaviest currently known being element 118 (oganesson).

1.1.1.1 Masses and binding energies

In contrast to materials in our everyday world, the mass of a given nucleus does not simply correspond to the sum of the masses of its component parts, in this case,

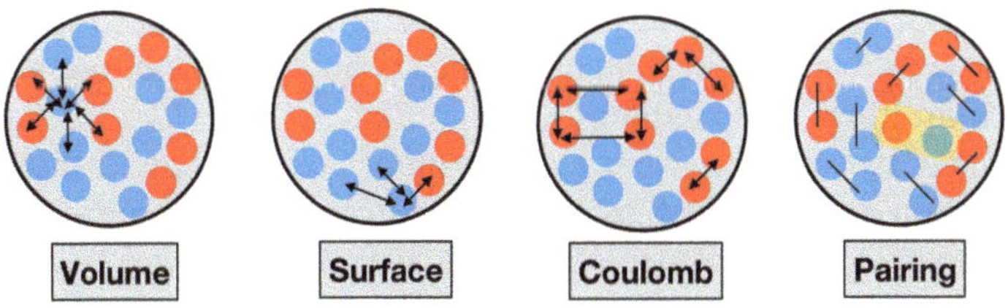

Figure 1.1. Schematic representation of four of the terms in the semi-empirical mass formula (equation (1.2)). Protons are shown in red and neutrons in blue. From left to right: the volume term which represents the fact that the strong nuclear force acts most effectively on nearest neighbours; the surface term which corrects for nucleons at the surface having fewer neighbours; the Coulomb term which represents repulsion between protons (in this schematic only a few of the relevant repulsions are shown); and the pairing term which reflects that like nucleons pair off (the unpaired nucleons in this example are highlighted in yellow). (Reproduced with permission from [67]. Copyright 2020 IOP Publishing.)

protons and neutrons. Rather, some of the mass is converted into the binding energy of the system, the latter varying from isotope to isotope. The mass of a nucleus, m_A with atomic mass, A, is given by

$$m_A = Zm_p + Nm_n - \frac{E_B}{c^2}, \tag{1.1}$$

where E_B is the binding energy, and m_p and m_n are the masses of the proton and neutron, respectively. Z is the atomic number (or number of protons) and N is the number of neutrons present in the nucleus. Extensive data exist on nuclear masses both for stable isotopes and for a growing number of unstable nuclei; such data are tabulated by the Atomic Mass Data Center (AMDC). The reader is directed to these data to explore for themselves trends in nuclear masses and binding energies. A fascinating aspect of atomic masses is that while it is very challenging to measure masses to high precision in our everyday world, it is readily possible to measure the masses of stable isotopes to extremely high precision, e.g. 1 part in 10^8. Relevant mass measurement techniques include the use of Penning traps and various approaches using storage rings (see chapter 6 of [1]).

The semi-empirical mass formula has long been used to derive estimates of E_B as a function of A and Z, where

$$E_B = a_V A - a_S A^{2/3} - a_c \frac{Z(Z-1)}{A^{1/3}} - a_{\text{asym}} \frac{(A-2Z)^2}{A} \pm \delta(A, Z). \tag{1.2}$$

Here, $a_V A$ is the volume term, $-a_S A^{2/3}$ is the surface term, $a_c \frac{Z(Z-1)}{A^{1/3}}$ is the Coulomb term, $-a_{\text{asym}} \frac{(A-2Z)^2}{A}$ is the asymmetry term and $\delta(A, Z)$ is the pairing term. The physical origin of some of these terms is presented in figure 1.1. The constants in these terms are conventionally fixed by fitting to data.

The balance of nuclear forces gives rise to the trend in binding energy per nucleon as a function of atomic mass, A, shown in figure 1.2. There are several interesting features of this plot. Notably, the function reaches a maximum around $A = 60$ and decreases both towards light nuclei and towards the heaviest. These trends imply that energy may be gained by splitting a heavy nucleus into smaller parts, i.e. the fission mechanism (see section 1.1.3) and in fusing light nuclei to produce heavier systems—a process common in stars and which is hoped to be harnessed as a future energy source on Earth (see chapter 2).

More detailed aspects of nuclear stability may be investigated through plotting particle separation energies, such as the one-neutron separation energy (the energy required to remove a single neutron from a nucleus), defined as

$$S_n = [-M(A, Z, N) + M(A-1, Z, N-1) + m_n]. \tag{1.3}$$

Exploring such trends in S_n for isotopic chains as a function of mass shows the importance of the pairing interaction in nuclear stability (see figure 1.3) which accounts for the staggering in separation energy between even-A and odd-A isotopes. Moreover, discontinuities in S_n for the calcium isotopes at $N = 20$ and $N = 28$ start to emerge, this effect becoming clearer when the two-neutron separation energy, S_{2n}, is plotted (see figure 1.4) which serves to remove the odd–even mass

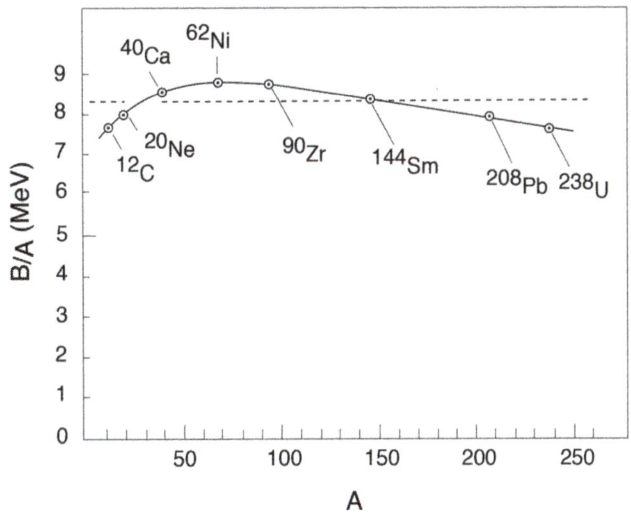

Figure 1.2. Binding energy per nucleon as a function of atomic mass, A. The location of a few specific nuclei are indicated on the diagram as a guide to the eye. (Reproduced with permission [68]. Copyright 2010 World Scientific.)

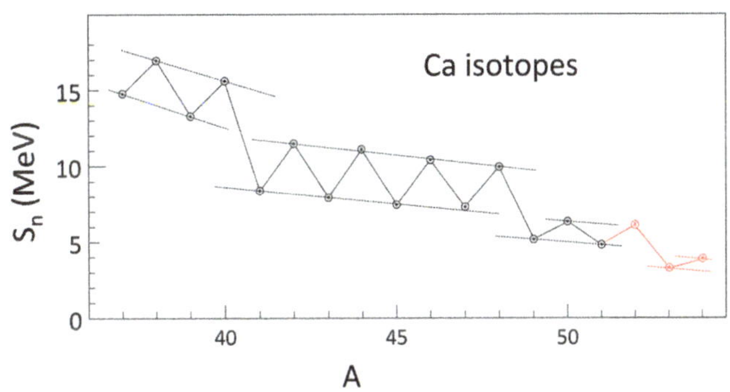

Figure 1.3. One-neutron separation energy, S_n, as a function of atomic mass, A for the calcium isotopes. Odd–even staggering shows the importance of the pairing interaction in nuclear stability. (Reproduced with permission from [68]. Copyright 2010 World Scientific.)

staggering. The observed discontinuities are understood as reflecting enhanced stability associated with these specific neutron numbers leading to them being denoted historically as *magic numbers*. Indeed, there is evidence from a wide range of experimental observables that supports such increased stability at the magic numbers, $N = Z = 2, 8, 20, 28, 50, 82$ and $N = 126$. The necessity to account for the occurrence of these magic numbers contributed to the development of the nuclear shell model in the late 1940s [4–6].

1.1.1.2 Nuclear radii and shape

The nuclear charge distribution was probed historically using electron scattering, electrons being chosen as they are effectively a point particle and have no internal

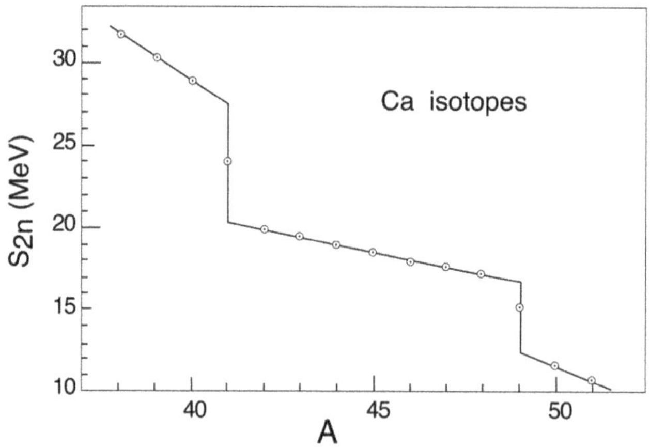

Figure 1.4. Two-neutron separation energy, S_{2n}, as a function of atomic mass, A for the calcium isotopes. The pronounced jumps at $A = 40$ and $A = 48$ support the concept of magic numbers for neutrons at $N = 20$ and $N = 28$. (Reproduced with permission from [68]. Copyright 2010 World Scientific.)

structure of their own. Early measurements showed the scaling of the nuclear radius as a function of atomic mass, A, followed the approximate relationship

$$R = r_0 A^{1/3}, \tag{1.4}$$

where r_0 is typically 1.2 fm $\equiv 1.2 \times 10^{-15}$ m. Importantly, the details of such scattering measurements showed that the nuclear radius could not be sharply defined because the nuclear surface was somehow fuzzy. The conventional form of the nuclear charge distribution takes account of this fuzziness, by introducing a surface diffuseness term, a, leading to

$$\rho(r) = \frac{\rho_0}{1 + \exp\left(\dfrac{r - R}{a}\right)}, \tag{1.5}$$

where ρ_0 is the central nuclear density which is of the order of 2×10^{17} kg m^{-3}, $\rho(r)$ is the density at radius r, and R is the nuclear radius (see equation (1.4)).

Deviations from a spherically symmetric charge distribution in the nucleus are represented at lowest order by the electric quadrupole moment, Q. Data for electric quadrupole moments across the nuclear chart are presented in figure 1.5. Since the quadrupole moment is a parameter which is only meaningfully defined for states with $J \geqslant 1$ (where J is the total angular momentum), the data presented are only for odd-Z and odd-N nuclei whose ground state satisfies this criterion. Close to closed shells (magic numbers for N or Z) the quadrupole moments are small, but in between closed shells the quadrupole moments become large. A simple model of a single unpaired proton in a particular nuclear orbital cannot account for such large values. The conclusion is that the nucleus as a whole must be deformed. Moreover, the majority of quadrupole moments are positive (see figure 1.5) which means that the nucleus is prolate deformed or rugby ball shaped. Why nuclei prefer to be deformed in this specific manner, rather than the alternative

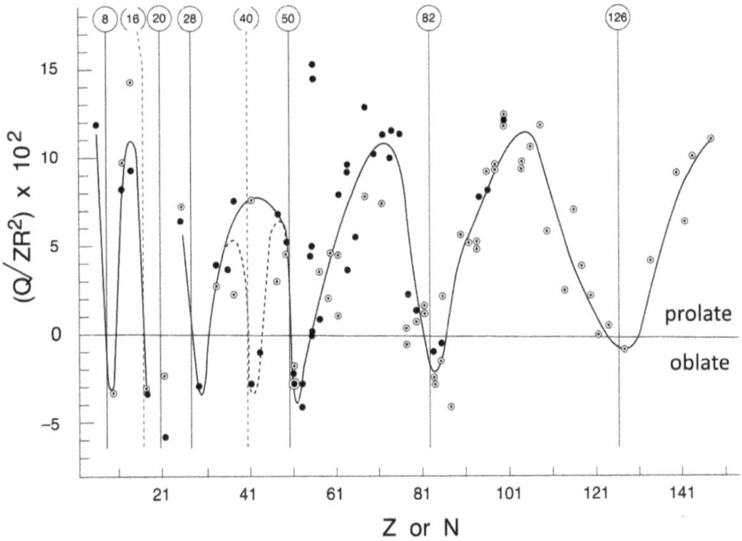

Figure 1.5. Electric quadrupole moments for odd-Z (solid circle) and odd-N (open circle) nuclei. Quadrupole moments have been normalized by a characteristic scaling factor of ZR^2 to allow all data to be represented on the same graph. The position of magic numbers (closed shells) for protons and neutrons are shown with the vertical lines labelled by the corresponding proton (or neutron) number. (Reproduced with permission from [68]. Copyright 2010 World Scientific.)

of oblate deformed (a flattened egg shape like the Earth), is unclear. Nevertheless, deformation appears ubiquitously across the nuclear chart with spherical nuclei standing out as unusual cases against this backdrop.

Exercises

1.1. Using masses from AMDC, plot S_n and S_{2n} for the iron isotopes ($Z = 26$). Identify any shell closures (magic numbers) for neutrons. Extrapolate the measured values to $S_{2n} = 0$ to determine which should be the last bound iron isotope.

1.2. Muon atomic spectroscopy is a technique used for measuring the charge radii of stable nuclei where a negative muon is added to the nucleus. Calculate the Bohr radius for a negative muon orbiting a heavy nucleus such as ^{152}Sm ($Z = 62$). Show that this Bohr radius is comparable to the nuclear radius expected from equation (1.4) and, hence, that muon orbitals are likely to be much more strongly influenced by the nuclear charge distribution than typical atomic electron orbitals.

1.3. A basic calculation for the quadrupole moment due to a proton in a certain shell model state with total angular momentum, j, is given by

$$\left\langle Q_{sp} \right\rangle = -\frac{2j - 1}{2(j + 1)} \langle r^2 \rangle,$$

where

$$\langle r^2 \rangle = \frac{3}{5} r_0^2 A^{2/3}.$$

Show that such an estimate cannot account for the majority of the measured quadrupole moments shown in figure 1.5

1.1.2 Nuclear excited states

The nucleus, in common with the atom, is a quantized system meaning that it cannot possess any given quantity of excitation energy but rather must occupy a limited set of excited states. Earlier volumes in this series discuss nuclear excited states in one of two broad categories: an independent-particle motion view where the pattern of excited states is dominated by the behaviour of one (or a few) individual nucleon(s) [3]; and a collective-motion view where the pattern of excited states is related to the component nucleons acting collectively [2]. There are abundant data in support of the latter view; indeed, rotational excitations in deformed nuclei are arguably the easiest class of excitations to identify. Rotational bands in nuclei broadly follow an energy dependence related to $J(J+1)$, where J is the angular momentum of the nucleus in that state; an example of such a rotational band in ^{238}U is shown on the right-hand side of figure 1.6. Due to symmetry, only states with even values of

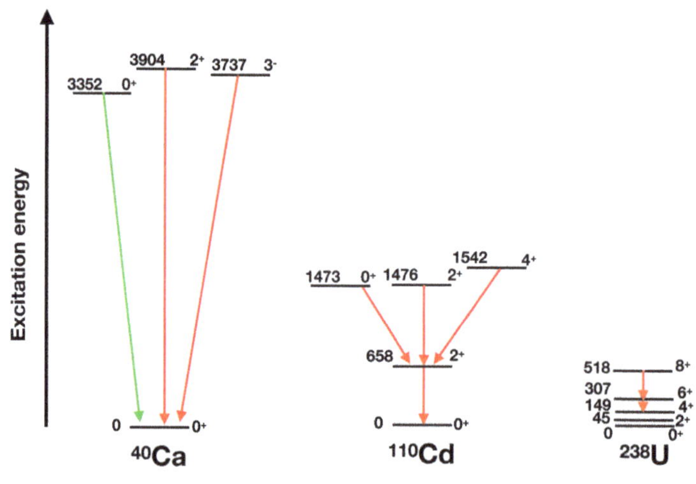

Figure 1.6. The first few excited states in some even–even nuclei: ^{40}Ca whose excited states are relatively high in energy, reflecting its nature as a doubly-magic ($Z = 20$, $N = 20$) nucleus, ^{110}Cd which has been suggested to be a vibrational nucleus, and ^{238}U whose pattern of excited states is typical of a deformed nucleus. The states are labelled by their excitation energy in keV and their spin-parity (J^π). By convention, the excitation energy of the 'ground state' is shown as 0. Observed electromagnetic transitions are indicated with the coloured arrows: the red arrows are gamma-ray transitions while the green arrow is the special case of internal pair transition since $0^+ \rightarrow 0^+$ transitions are forbidden to occur by gamma-ray emission (see section 1.1.3). Transitions between the low-lying states of ^{238}U are said to be highly converted, meaning that there is a high probability that they occur through emission of conversion electrons (see section 1.1.3). (Reproduced with permission from [67]. Copyright 2020 IOP Publishing.)

angular momentum appear in a rotational band built on the ground state in such an even–even nucleus. Moreover, the $J(J+1)$ energy dependence implies that the ratio of the excitation energies of the 4^+ and 2^+ states should be in the ratio $R_{42} = \frac{10}{3} = 3.333\ldots$. This ratio is therefore frequently used as a means of identifying rotational behaviour in nuclei; see chapter 2 of [1] for more details.

Exercises

1.4. Retrieve data on excited states in ^{24}Mg, ^{174}Yb and ^{242}Pu from NNDC. Identify the states comprising a rotational band, i.e. the $E \propto J(J+1)$ sequence built on the ground state. How closely do these states confirm to the expected $J(J+1)$ energy dependence?

1.1.2.1 Labelling excited states

Excited states of a nucleus are conventionally labelled according to their excitation energy, E_x, total angular momentum J and parity π (see figure 1.6). By convention, the ground state of a nucleus is defined to have an excitation energy of zero and the excitation energies of excited states are presented as positive values relative to this definition. However, it should be noted that stable nuclei will in reality have ground states with an absolute excitation energy which is negative.

It is quite common in nuclear structure to hear total angular momentum, J, described as 'spin', leading to the shorthand notation of spin-parity or J^π. Here, parity (π) reflects the symmetry of the system. The *parity* operator acts on a wavefunction $\psi(\vec{r})$ or potential $V(\vec{r})$ by reflecting the coordinate system used to describe it, i.e. in a Cartesian system $\vec{r} \rightarrow -\vec{r}$ where $\vec{r}$ is a vector with components $x \rightarrow -x$, $y \rightarrow -y$ and $z \rightarrow -z$. Potentials are generally invariant under the parity transformation, i.e. $V(-\vec{r}) = V(\vec{r})$, but wavefunctions may not be, i.e. $\psi(-\vec{r}) = \pm\psi(\vec{r})$. For cases where the parity operator causes no change of sign in the wavefunction, the wavefunction is said to have *even* or *positive* parity ($\pi = +$) and for cases where the sign changes the parity is *odd* or *negative* ($\pi = -$).

Since protons and neutrons both have an intrinsic half-integer spin, the total angular momentum of states in a given nucleus with an even number of protons and an even number of neutrons must be integer. In fact, symmetry considerations mean that the ground state of such even–even nuclei will always have $J^\pi = 0^+$ while their first excited state is nearly always (with some special exceptions) $J^\pi = 2^+$. Excited states in *odd-A* nuclei, those with an odd number of either protons or neutrons (but not both), will have a half-integer value for J. Excited states in nuclei with an odd number of both protons and neutrons, so-called *odd–odd* nuclei, must have integer values for J but there is not the same symmetry acting as in even–even nuclei and the ground state of such nuclei can take on various integer values of J^π.

A further labelling of excited states useful in elucidating aspects of nuclear structure is the *isospin* quantum number, t or T. The isospin labelling is related to the fact that the proton and neutron are two aspects of a common class of particle—the nucleon. Both protons and neutrons have isospin, $t = 1/2$ with projections of

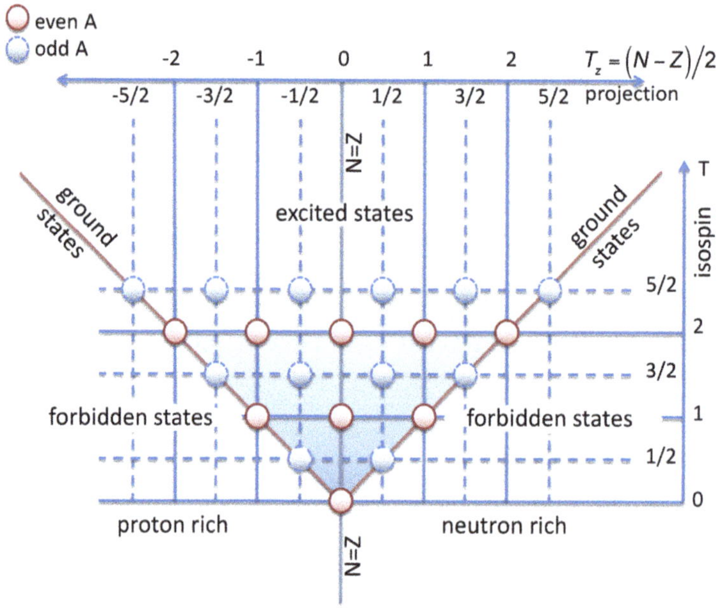

Figure 1.7. The isospin triangle illustrates certain allowed and forbidden combinations of isospin, T, and isospin projection, T_z. For example, an even–even nucleus with $N = Z$ would be expected to have a ground state and low-lying excited states with $T = 0$, $T_z = 0$. At somewhat higher energy, e.g. 10 MeV, states with $T = 1$, $T_z = 0$ would be expected to appear. Such states would be isobaric analogue states of the corresponding ground state and low-lying excited states in the neighbouring $T = 1$, $T_z = \pm 1$ nuclei. (Reproduced from [7]. CC-BY-4.0.)

$t_z = -1/2$ for the proton and $t_z = +1/2$ for the neutron. The isospin, T, and isospin projection, T_z of ground states and excited states in nuclei can be evaluated by taking account of the total number of protons and neutrons and their coupling (see figure 1.7). Isospin symmetry has particular utility in organizing nuclear excited states since isospin mixing is in most cases negligible.

Exercises

1.5. Use ENSDF to look up data for excited states in ^{24}Na, ^{24}Mg and ^{24}Al. Identify the high-lying (around 10 MeV in excitation energy) isobaric analogue ($T = 1$) states in ^{24}Mg and compare these with the low-lying $T = 1$ states in ^{24}Na and ^{24}Al. How well is isospin symmetry respected in this $A = 24$, $T = 1$ triplet system?

1.1.2.2 Level density

As can be seen in the examples provided in figure 1.6, the level spacing in even–even nuclei at low excitation energy can be relatively large, e.g. 100 keV–1 MeV. Level density in neighbouring isotopes can be quite different, however. If we consider some specific even–even nucleus, then its neighbouring odd-A nucleus would be expected

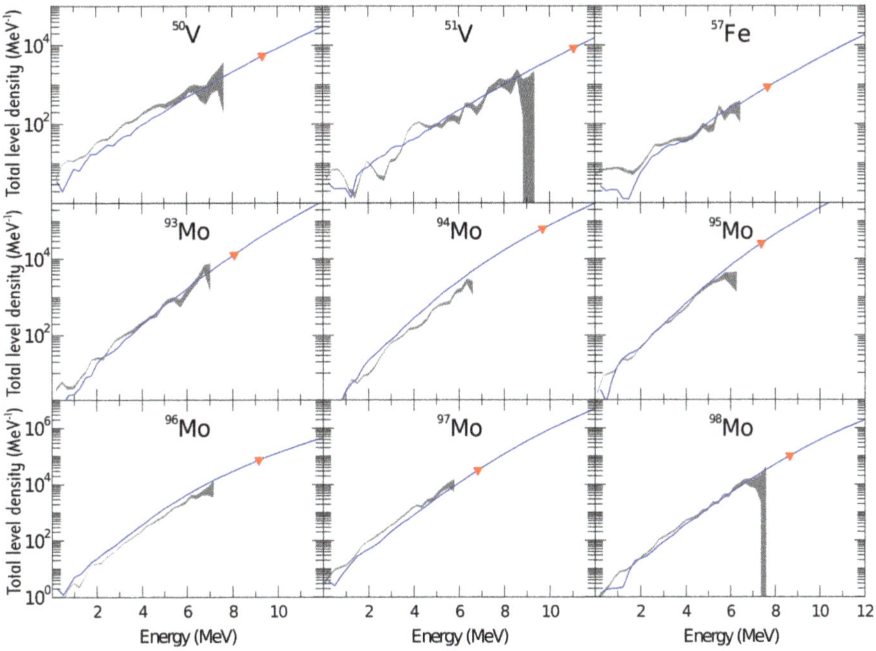

Figure 1.8. Total level density as a function of excitation energy in a range of different nuclei. The grey band represents data extracted from experiment while the blue curve represents a theoretical model. (Reproduced with permission from [8]. Copyright 2019 Elsevier.)

to have relatively more excited states per energy range, while the neighbouring odd–odd nucleus would have a significantly higher level density. The latter is due to the many possible permutations of coupling the angular momentum of the odd proton and odd neutron. As a function of excitation energy, the number of excited states in all nuclei, irrespective of whether they are even–even, odd-A etc, increases exponentially (see figure 1.8). This latter effect proves quite important for various nuclear reaction models because as reactions access regions of high level density the characteristics of individual states become less important.

Exercises

1.6, Look up data on excited states in ENSDF for ^{70}Se, ^{71}Br and ^{72}Br. The latter correspond to systems with one proton more than ^{70}Se, and one proton and one neutron more than ^{70}Se, respectively. Compare the number of excited states known below 2 MeV.

1.1.2.3 Lifetime of excited states

Nuclear decay is a *stochastic process* meaning that we cannot predict when a particular nucleus will decay but we can define a meaningful probability for the likelihood of such a decay in an interval of time. The time dependence of radioactive decay is given by

$$N(t) = N_0 e^{-\lambda t}, \tag{1.6}$$

where N_0 is the initial number of nuclei in a given state (e.g. excited state or the ground state of an unstable nucleus), $N(t)$ is the number remaining in that state at a subsequent time, t, and λ is the decay constant (in units of s^{-1}). The lifetime of a decay, τ, is $1/\lambda$ and corresponds to the time taken for the activity of the sample to reduce to $1/e$ of the initial value. Since it is perhaps more intuitive to think about the sample's activity reducing to half the initial value, it is common to discuss half-life, $t_{1/2}$, rather than lifetime where $t_{1/2} = 0.693\tau$. Lifetime and half-life are used rather interchangeably in nuclear physics, although the former is more commonly used for the decay of short-lived nuclear excited states and the latter for the decay of nuclear ground states.

Nuclear excited states exhibit a broad range of lifetimes (or half-lives). *Bound excited states* of nuclei decay principally by gamma-ray emission with internal conversion as a competing decay mode, which is important in certain cases such as heavy nuclei, e.g. the actinide region (see section 1.1.3). The probability of gamma decay from a bound excited state, which may be by multiple branches, will essentially define the lifetime of the state. Typical excited state lifetimes are in the fs (10^{-15} s) to ps (10^{-12} s) range. Some nuclei possess excited states which are significantly more long-lived; such metastable states are referred to as *isomers*. A lower limit on isomer lifetime is not so well defined (and is often application specific) but the definition of an isomer is typically taken to mean a state with a lifetime exceeding a few nanoseconds; in certain extreme cases, gamma-decaying isomers may even have lifetimes as long as several years. The general condition for a state to be isomeric is for its gamma decay to available lower energy states within the same nucleus to be strongly hindered. There are several ways in which such hindrance might occur. A common reason is that there is a large angular momentum difference between the isomer and daughter state(s) that it might decay into. Figure 1.9 illustrates this case and some other cases where nuclear isomerism occurs.

In all nuclei, once sufficient excitation energy is put into the system, it will exceed one or more of the particle separation energies, e.g. the neutron separation energy, S_n (see definition in equation (1.3))[1]. States above the particle separation thresholds are described as *unbound states*. As excitation energy increases still further beyond the particle separation threshold, the probability for particle emission will become overwhelming and gamma decay of such states becomes a very low probability decay branch. Having more open decay branches implies that the lifetime of the excited state becomes shorter and shorter. In this regime, it is a useful convention to switch over to discussing the (energy) widths of unbound states rather than their lifetime, where their total width, Γ_{TOT} connects to the lifetime via the Heisenberg uncertainty relation,

[1] In fact, some very exotic nuclei are unbound against particle emission even in their ground state.

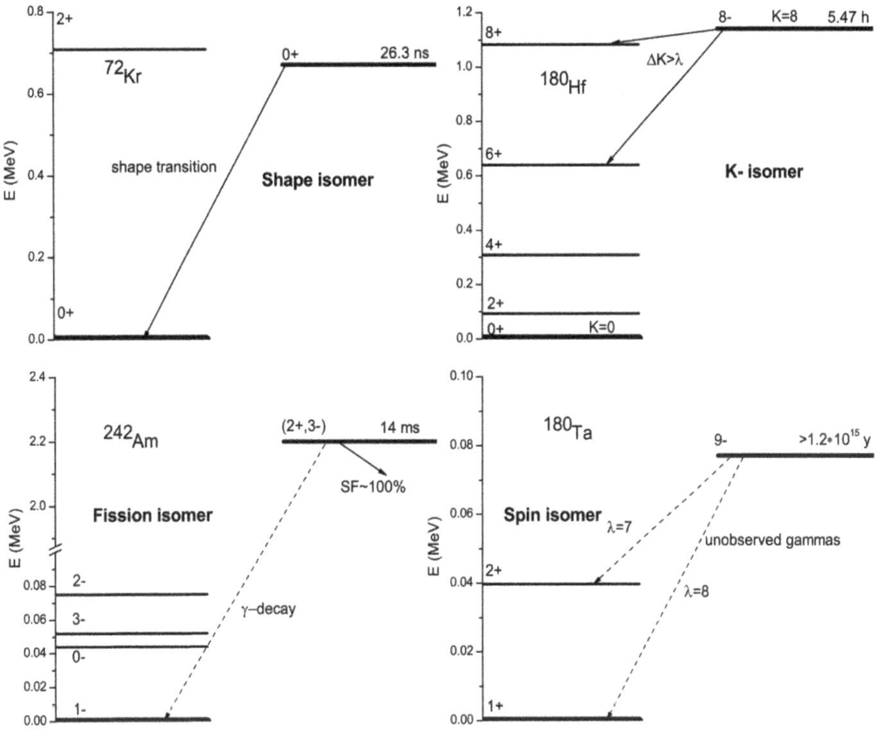

Figure 1.9. Four examples of isomeric states labelled with their respective half-life. Top left: the first excited state (0^+) in ^{72}Kr which appears, unusually, below the first 2^+ state due to shape coexistence in this nucleus. The $0^+ \rightarrow 0^+$ decay is slow because it can occur only through internal conversion. Top right: a so-called K isomer where the decay is retarded due to the nuclear structure of the isomer which has the total angular momentum vector oriented in a very different direction to that of the states which it can decay to. The interested reader will find more on the origins of such K isomerism in section 3.4.1 of [1]. Bottom left: a fission isomer in ^{242}Am which is a highly deformed excited state that can fission more readily than undergo energetically allowed gamma decay. Bottom right: a spin isomer in ^{180}Ta which cannot gamma decay because the associated angular momentum change implies a very high multipolarity for the gamma-rays, e.g. E7 (see section 1.1.3 where multipolarity is defined) which leads to a highly retarded decay. (Reproduced with permission from [69]. Copyright 2015 Elsevier.)

$$\Delta E \Delta t \sim \hbar. \tag{1.7}$$

The total width of a state may be expressed as the sum of the partial widths for the various possible decay branches, e.g. for a state unbound to both proton and neutron emission, the total width would be expressed as

$$\Gamma_{\mathrm{TOT}} = \Gamma_\gamma + \Gamma_n + \Gamma_p, \tag{1.8}$$

where Γ_γ is the partial width for gamma emission, Γ_n is the partial width for neutron emission and Γ_p is the partial width for proton emission.

Exercises

1.7. On the basis of equation (1.7), calculate the total width of gamma-decaying states with lifetimes of 1 ps and 1 fs.

1.1.3 Nuclear decay modes

Ground states of unstable nuclei (and long-lived isomeric states) may decay by a range of different decay modes including alpha and beta decay as well as spontaneous fission. In general, short-lived nuclear excited states decay by gamma-ray emission (or internal conversion). Here, we summarize the characteristics of the most common decay modes in turn.

1.1.3.1 Alpha decay

Many radioactive isotopes of heavy elements such as the actinides decay by emitting alpha particles ($^4_2$He nuclei) in order to reach a more stable configuration. In this process, the atomic number of the nucleus decreases by two and the atomic mass by four, e.g.

$$^{238}_{92}\mathrm{U} \rightarrow {}^{234}_{90}\mathrm{Th} + {}^{4}_{2}\mathrm{He}. \tag{1.9}$$

A striking feature of alpha decay is that the α particle is not itself a fundamental particle; the fact that it is spontaneously emitted as such a subunit from heavy nuclei reflects the high stability of the $^4_2$He nucleus. The physics of alpha particle emission is governed by *quantum tunnelling*. This means that a pre-formed alpha particle within the nucleus does not have sufficient kinetic energy, in a classical sense, to overcome the Coulomb barrier and escape from the nucleus. Instead, the decay process takes place by tunnelling through the barrier (see figure 1.10).

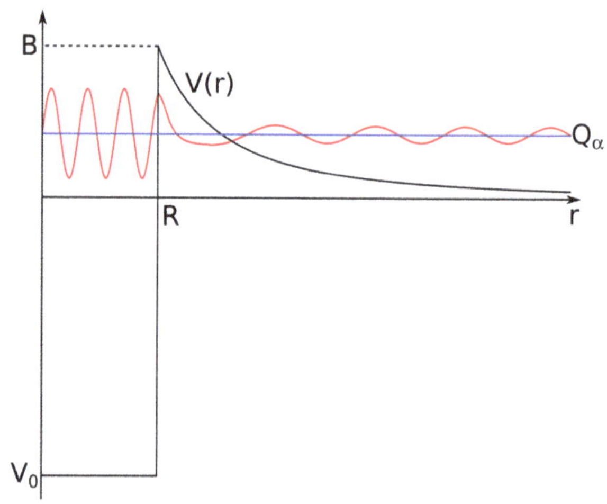

Figure 1.10. General schematic of alpha tunnelling showing an idealized nuclear square well potential with depth, V_0 and the Coulomb barrier height, B. The potential falls off as a function of radius, $V(r)$, outside the nuclear radius, R. The Q value for the decay, Q_α, is indicated. (Reproduced with permission from [67]. Copyright 2020 IOP Publishing.)

The probability of alpha decay depends strongly on the Q-value for the decay which can be calculated from the respective masses

$$M({}^{A}_{Z}X_{N}) = M({}^{A-4}_{Z-2}X_{N-2}) + M({}^{4}_{2}\mathrm{He}) + Q, \tag{1.10}$$

where, for example, $M({}^{A}_{Z}X_{N})$ is the atomic mass of isotope X, which has atomic mass, $A = Z + N$, and Q is in atomic mass units ($1u$: $=931.49$ MeV/c^2). The concept of Q-value will appear later in the context of nuclear reactions. For radioactive decay, Q-values are only positive, i.e. energy is only released. For reactions, Q-value can be positive or negative. Where Q is negative, it represents the amount of energy required to be input for a reaction to proceed and, when positive, the energy liberated by the reaction, paralleling the case of radioactive decay.

Since alpha decay is a two-body process, the recoil of the heavy nucleus must be taken into account in transforming from the Q-value of the decay to the kinetic energy of the emitted alpha particle, T_α via the relationship

$$T_\alpha = \frac{Q}{1 + \frac{m_\alpha}{m_d}}, \tag{1.11}$$

where m_α is the atomic mass of the alpha particle and m_d is the atomic mass of the daughter nucleus. Typical values of T_α are in the range 3–10 MeV. Depending on the details of the nuclear structure of the daughter nucleus, alpha decay may proceed not only to the ground state of the daughter but also to excited states leading to alpha decay *fine structure.*

The so-called *Geiger–Nuttall* relationship is an empirically derived correlation between alpha decay half-lives and the Q-value of the decay, given by

$$\log T_{1/2} = A(Z)Q^{-1/2} + B(Z), \tag{1.12}$$

where $A(Z)$ and $B(Z)$ are functions of the atomic number, Z. Alpha decay half-lives vary over a wide range from fractions of a second in exotic nuclei to millions/billions of years. In fact, the alpha decay of long-lived heavy (actinide) nuclei proves important in the handling of spent nuclear fuel since the alpha particles emitted may induce neutron-emitting reactions in surrounding material. This issue is discussed in more detail in section 5.3.1.

1.1.3.2 β^- decay

Alpha decay is a decay mode mostly associated only with the heaviest nuclei. A far more common decay mode for nuclear ground states and long-lived isomeric states is beta decay where, for a neutron-rich nucleus, a neutron transforms into a proton via the weak interaction as a means of achieving greater stability for the system as a whole. Very early in the history of nuclear physics, Chadwick [9] had observed that beta decay spectra are continuous, in contrast to alpha particle and gamma-ray spectra that consist of individual discrete lines. Pauli proposed that a neutral particle of negligible mass, later called the neutrino, existed as a solution to this observation. If emitted with an electron in beta decay it would explain the nature of the beta decay spectrum because the energy released would be shared by the two outgoing particles and recoiling nucleus[2]. Indeed, we now understand that a pair of leptons (an electron—commonly referred to as a β^- particle—and an anti-electron-neutrino—$\bar{\nu}_e$) must be created in order to conserve charge and lepton number. A well-known example of such decay is that of ^{14}C into ^{14}N via the following process

$$^{14}_{6}\mathrm{C} \rightarrow {}^{14}_{7}\mathrm{N} + \mathrm{e}^- + \bar{\nu}_e \tag{1.13}$$

with a half-life of 5730 years, this decay and its relatively long half-life being the basis of the carbon dating technique. Similarly, a free neutron may undergo decay into a proton

$$\mathrm{n} \rightarrow \mathrm{p} + \mathrm{e}^- + \bar{\nu}_e \tag{1.14}$$

with a lifetime of around 880 s.[3] It is interesting to reflect that neutrons bound in nuclei, in general, do not decay so rapidly if at all. Evidently, there is a strong dependence of the probability of beta decay on the nuclear medium and the details of nuclear structure, which we will come to shortly (see section 1.1.3).

As with alpha decay, the energetics of the decay are important in defining the probability of a given beta decay. The Q-value, again in atomic mass units, for β^- decay is given by

$$M({}^{A}_{Z}X_N) = M({}^{A}_{Z+1}X_{N-1}) + Q(\beta^-). \tag{1.15}$$

If an isotope is particularly neutron-rich it may undergo successive β^- decays until it reaches a stable configuration (see an example for the $A = 131$ isobaric chain in figure 1.11). Indeed, such multi-step decays are relatively common in the case of so-called *fission fragments* produced in spontaneous or particle-induced fission (see section 1.1.3).

As mentioned above, a consequence of the β^- decay process where both an electron and an antineutrino are emitted is that the energy of the emitted electrons (beta particles) is not discrete but, rather, is continuous and follows a so-called Fermi–Kurie distribution (see figure 1.12). Such electron energy distributions may readily be measured and codes such as BetaShape exist which can be used to calculate them. In contrast, detection of the emitted antineutrino is extremely challenging because such particles hardly interact with matter at all. Nevertheless, antineutrinos can be detected and this has been suggested as a means of remotely monitoring the fuelling of fission reactors; we cover this topic in section 4.2.2.

1.1.3.3 β^+ decay and electron capture (EC) decay

Isotopes may also be unstable due to an excess of protons. In this case, a proton within the nucleus may transform into a neutron mediated by the weak interaction. As with β^- decay, due to conservation laws, a pair of leptons (a positron and an

[2] Although the existence of neutrinos was not confirmed until they were observed by Reines and Cowan [10, 11], Fermi [12] had already incorporated it into his theory of beta decay based on the same formalism as electromagnetism.

[3] Intriguingly, there is some controversy on this point since the two main measurement techniques for the lifetime of the neutron do not give consistent results—the so-called neutron lifetime puzzle.

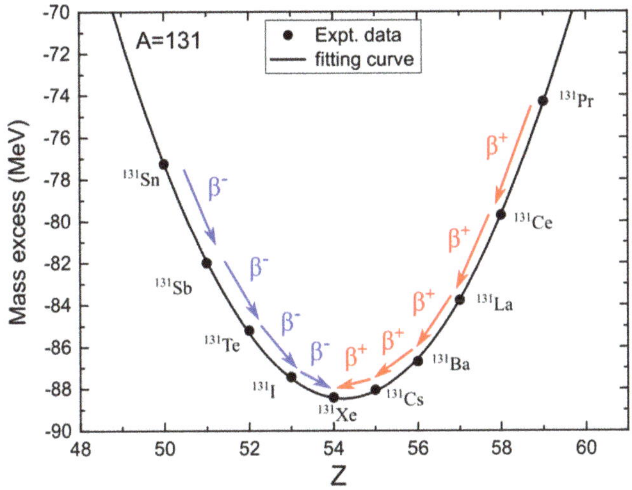

Figure 1.11. The mass parabola for the $A = 131$ isobaric chain: the mass excess is plotted as a function of Z. Observed beta decays are labelled with blue arrows (β^-) and red arrows (β^+/EC). One can observe pictorially how beta decay Q-values decrease on approaching stability, implying longer half-lives for the respective decays. (Reproduced with permission from [67]. Copyright 2020 IOP Publishing.)

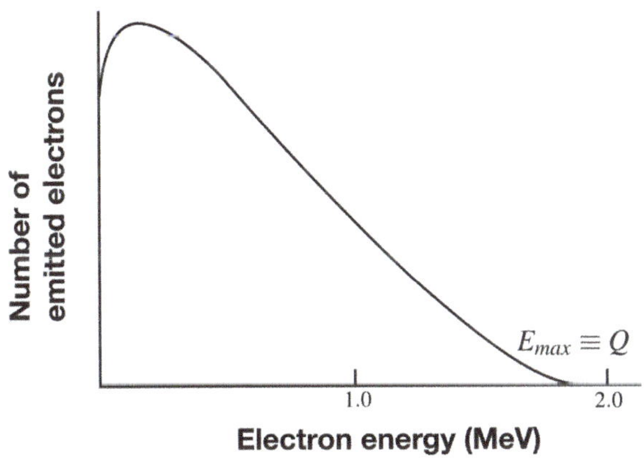

Figure 1.12. A schematic of the Fermi–Kurie distribution of emitted electron energies in β^- decay. The energies follow a continuous distribution up to an endpoint given by the Q-value. (Reproduced with permission from [67]. Copyright 2020 IOP Publishing.)

electron-neutrino, ν_e) is produced in the process. The decaying nucleus decreases its atomic number by one, but its mass number remains unchanged, namely

$$^{12}_{7}\mathrm{N} \rightarrow {}^{12}_{6}\mathrm{C} + \mathrm{e}^+ + \nu_e. \tag{1.16}$$

The Q-value for β^+ decay is given by

$$M(^A_ZX_N) = M(^A_{Z-1}X_{N+1}) + 2m_e + Q(\beta^+), \tag{1.17}$$

where m_e is the rest mass of the electron. The factor of 2 is due to the unbinding of one atomic electron from the atom following the decay of one of the protons present in the decaying nucleus. Equation (1.17) assumes that the decaying nucleus is part of a neutral atom; for the case of an ionized initial nucleus, the factor of 2 would disappear.

A key difference between β^+ and β^- decay is that there is a further possible decay branch in competition. This is decay by electron capture, where the nucleus captures an atomic electron to mediate the decay. Here, an example electron capture (EC) decay would be

$$^{12}_{7}\mathrm{N} + \mathrm{e}^- \rightarrow {}^{12}_{6}\mathrm{C} + \nu_e. \tag{1.18}$$

Since there are two bodies in the exit channel rather than three in the case of β^- or β^+ decay, the neutrinos will be emitted with discrete energies. The relevant Q-value is given by

$$M({}^{A}_{Z}X_N) = M({}^{A}_{Z-1}X_{N+1}) + Q(\mathrm{EC}). \tag{1.19}$$

As can be seen from a comparison of equation (1.19) with equation (1.17), there is an additional energy cost of $2m_e$ in β^+ decay to generate the emitted positron leading to a certain minimum Q-value before β^+ can occur. This means that EC is the only process possible for $Q(\mathrm{EC}) < 1.022$ MeV. Above that critical value, β^+ becomes increasingly important and for large values of Q(EC), it is by far the dominant process. As illustrated in figure 1.11, a sufficiently proton-rich nucleus may decay by a succession of β^+/EC decays to reach stability.

1.1.3.4 States populated in beta decay

Here, we describe the physics of beta decay in a little more detail since we will be returning to beta decay in Chapter 3 as the main contributor to decay heat in fission reactors. Indeed, we will see how details of beta decay constitute a specific source of nuclear data uncertainty. The reader who seeks a more detailed description of beta decay can find it in, for example, [13, 14].

Irrespective of whether a nucleus decays by β^- or β^+/EC decay, the decay may, in general, populate a number of excited states and/or the ground state in the daughter nucleus. How many states are populated depends on a number of factors. The most important is the Q-value—the difference in the nuclear masses of the two nuclear ground states (see equation (1.15)). Here, depending on the mode of decay, the relevant Q-value could be either $Q(\beta^-)$ or Q(EC). The larger the Q-value, the higher the energy of the possible final states that can be populated. Naturally, the emitted electron or positron will be attracted or repelled by the nuclear charge. Their wave functions are enhanced or suppressed while they are close to the nucleus. These two effects can be calculated. Their joint effect is called the Fermi Integral (f) which can be found in tabulated form or calculated employing web-based tools such as LogFT. The third main factor affecting the transition rate is the change in nuclear structure. This is given by the matrix element M_B extracted from

$$<\psi_f\left|\hat{O}\right|\psi_i>, \tag{1.20}$$

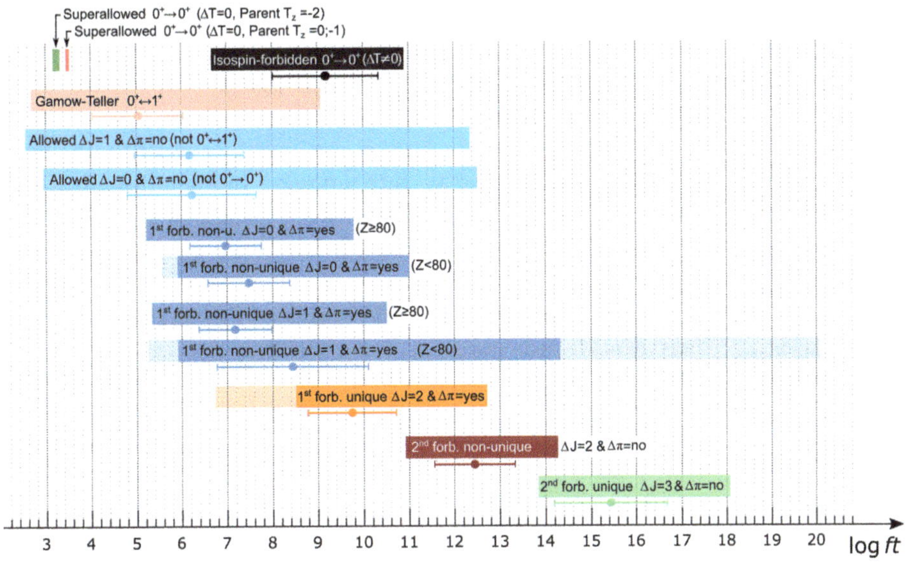

Figure 1.13. Distribution of $\log ft$ values for different classes of beta decay including Fermi super-allowed, allowed Fermi and Gamow–Teller decays and various classes of forbidden transition. For a brief introduction to the meaning of these terms, see the text for further details. (Reproduced with permission from [15]. Copyright 2023 The Authors. Published by Elsevier Inc. CC-BY-NC-ND 4.0.)

where ψ_f and ψ_i are the wave functions of the final and initial states and $\hat{O}$ is the operator responsible for the beta transition. M_B cannot be measured directly. It has to be obtained from the measurement of the beta transition probability; this is usually denoted by B and is the square of M_B. In practice, one normally uses what is called the ft value where the Fermi integral is combined with t the partial half-life[4] for the transition to an individual state in the daughter nucleus. It turns out that M_B^2 is proportional to $1/ft$. The measured values that are found in nuclei cover many orders of magnitude. Accordingly, they are always quoted in terms of $\log ft$ (see figure 1.13). The measured $\log ft$ values were empirically classified, with the fastest transitions being called allowed transitions and others with larger $\log ft$ said to be forbidden. It is very important to note that this nomenclature was introduced without any understanding at the time of what determines the transition rates. In reality, the transitions labelled as forbidden actually do occur—they are simply slower than allowed transitions.

Fermi's theory of beta decay provided an explanation for the wide variation in beta decay probability. In essence, his theory was based on analogy with the electromagnetic interaction. There are five possible operators with the correct mathematical form to satisfy the theory. They are the vector (V), scalar (S), pseudoscalar (P), axial vector (A) and tensor (T) interactions. It took two decades

[4] The partial half-life (t) is defined as the ratio between the total half-life of the decay ($T_{1/2}$) and the probability of the particular process or branch, represented here by P. So, it can be calculated as $t = \frac{T_{1/2}}{P}$.

before it was clear whether Fermi's assumption of V or S, or the T or A introduced by Gamow and Teller were correct. It was found that neither charge conjugation (C) [16] nor parity (P) [17] are conserved in beta decay although the combination CP is; this meant that V and A were confirmed as the relevant operators.

Fermi assumed that the interaction takes place at a point. In essence, this means that the electron and neutrino carry away zero angular momentum ($L = 0$). In modern theory, beta decay is explained in terms of the exchange of $W^{+/-}$ bosons with a mass of ~80 GeV/c^2. Thus, Fermi's assumption that the interaction occurs at a point and the change in angular momentum is zero is justified.

If we define S as the vector sum of the electron and neutrino intrinsic spins, both of which are 1/2, then $L = 0$ and $S = 0$ or 1 are the possible combinations for allowed transitions. When the spins are antiparallel $S = 0$, we have a Fermi transition ($\Delta J = 0$, where $J = L + S$, is the total spin) since it corresponds to Fermi's first calculations. The alternative is that $S = 1$ so then $\Delta J = 0$ or ± 1.[5] This is called a Gamow–Teller transition.

The forbidden transitions have $L \neq 0$ and are suppressed when compared with allowed transitions. There are degrees of forbiddenness with the transition probability decreasing as we move from first to second to third forbidden, etc. In reality, allowed and forbidden transitions have an overlap in logft values (see figure 1.13). We will not go into this in detail but it depends a lot on the Q-value. As it increases, the relative probability increases for forbidden transitions, but it depends heavily on the available single particle orbitals.

In terms of isospin T, each nucleus has $T_z = (N - Z)/2$ (see section 1.1.2). The only unhindered Fermi transitions are the so-called super-allowed transitions, where $\Delta T = 0$. These transitions occur between isobaric analogue states (IASs)[6]. They can only occur when the IAS lies within the Q-value window and are relevant to determining the value of the constant that determines the strength of the weak interaction.

1.1.3.5 Beta-delayed particle emission

Depending on the Q-value for the decay, it may be possible for beta decay to take place to excited states which are unbound against particle emission. This can lead to processes such as beta-delayed proton (β_p) emission. Figure 1.14 shows a schematic of beta-delayed proton emission while figure 1.15 shows a schematic of possible decays of an exotic, proton-rich nucleus where the states populated may decay by multiple proton emission and even alpha emission. The process of beta-delayed neutron (β_n) emission turns out to be very important in the regulation of fission reactors and will be returned to, in some detail, later in section 3.4.

[5] It should be noted that a 0^- to 0^- transition of this kind is not possible.

[6] IASs are the states found in neighbouring isobars (nuclei with the same mass number A) that have the same total angular momentum, parity and isospin. If we assume that the nuclear force is approximately charge-independent, then isospin is a good quantum number. As a consequence, IASs in different nuclei should have a very similar nuclear structure.

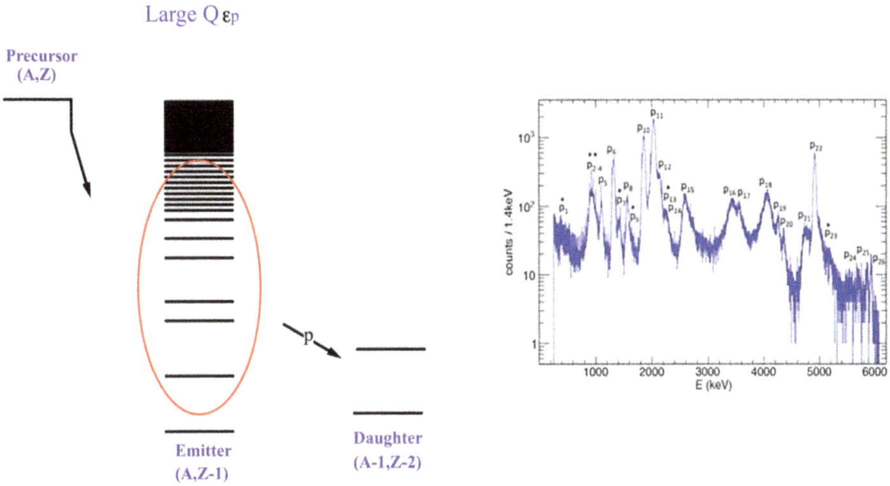

Figure 1.14. A schematic of the decay of an isotope which has a large value of Q(EC) allowing some portion of the beta decay to proceed from the precursor nucleus to proton-unbound states in the emitter nucleus which emit protons to reach states in the daughter nucleus. A typical beta-delayed proton spectrum measured experimentally is shown on the right-hand side of the figure. (Reproduced with permission from [18]. Copyright 2020 Elsevier.)

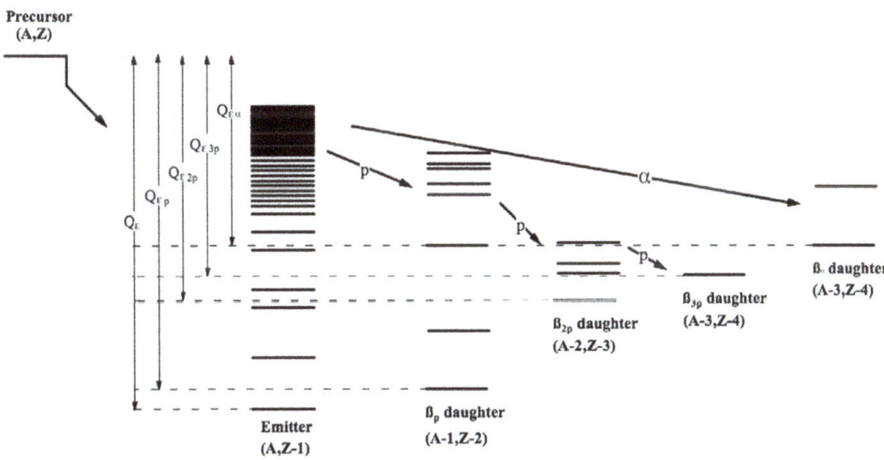

Figure 1.15. A schematic of the decay of an exotic proton-rich isotope which can beta decay to a wide range of excited states that in turn decay by emitting one (or multiple) protons and/or alpha particles. Where energy differences are labelled on the figure Q_ϵ, this corresponds to Q(EC) in the notation used in the text. (Reproduced with permission from [18]. Copyright 2020 Elsevier.)

1.1.3.6 Gamma decay

Nuclei which are unstable due to an excess of excitation energy or angular momentum may reach a more stable configuration through the emission of γ-rays. Frequently, such decay may involve a cascade of gamma-rays emitted through a sequence of excited states down to the ground state or a long-lived metastable state or isomer[7].

While gamma-ray emission is, in principle, a two-body process, in practice, in particular for heavy nuclei, the nuclear recoil is negligible and the energy of an emitted γ-ray is extremely close to the excitation energy difference between the initial and final nuclear level. Gamma-ray emission may lead to angular momentum being carried away from the system. For a transition from a state with angular momentum vector $\overrightarrow{J_i}$ to a state with vector $\overrightarrow{J_f}$, the angular momentum vector of the photon $\overrightarrow{L}$ must complete the vector triangle[8]:

$$\overrightarrow{J_i} = \overrightarrow{L} + \overrightarrow{J_f}. \tag{1.21}$$

This implies that a range of values of L may contribute to the decay where the largest possible value of L is $|J_i + J_f|$ while the smallest possible value of L is $|J_i - J_f|$. Gamma-rays also possess electric (E) or magnetic (M) character. This is related to the change in parity between the initial and final states mediated by the decay. The so-called *multipolarity* of the transition is presented as EL or ML e.g. M1, E2, E5 etc. Parity selection rules apply where

$$\pi(\text{ML}) = (-1)^{L+1} \tag{1.22}$$

and

$$\pi(\text{EL}) = (-1)^{L}. \tag{1.23}$$

In general, the higher the multipolarity, the less probable the decay mode, although it is quite common for transitions to exhibit a mixture of two predominant multipolarities, e.g. M1/E2, in which case the mixing of these multipolarities is defined by a mixing ratio, δ. The interested reader will find more on gamma-ray transition probabilities in chapter 2 of [1] which also introduces the concept of *Weisskopf estimates* which are the yardstick for defining relative transition strength.

1.1.3.7 Internal conversion

Internal conversion (IC) is a process which can compete with gamma-ray emission. Here, the excited nucleus interacts electromagnetically with one of its orbiting electrons and de-excites by transferring some or all of its excitation energy to the electron which is then emitted from the atom (see the right-hand side of figure 1.16). It should be noted that this emitted conversion electron is quite distinct from the electron emitted in β^- decay; indeed the atomic mass and Z of the nucleus remains unchanged after internal conversion in the same manner as in γ decay. The competition between IC and gamma-ray emission for a given transition is quantified by a so-called *conversion coefficient*, α_{ic} where

[7] It is important to note that, in distinction to the α and β emission processes, the atomic number and atomic mass of the emitting nucleus remain the same following γ-ray emission.

[8] A transition of the type $0^+ \rightarrow 0^+$ cannot proceed by gamma-ray emission because the photon has an intrinsic spin of $1\hbar$. Such a transition can only proceed by internal conversion or by an internal pair transition.

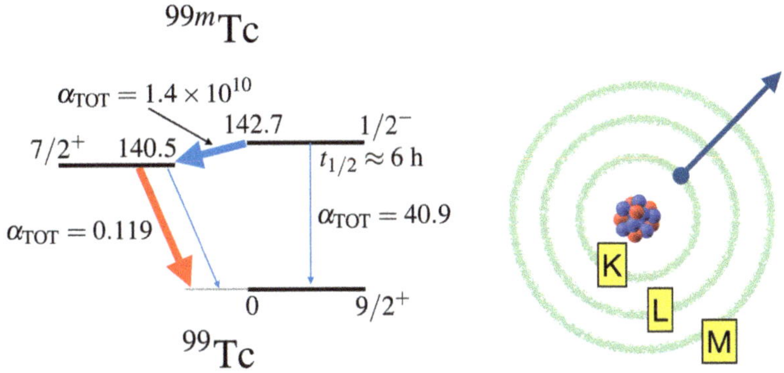

Figure 1.16. Left: A partial level scheme showing the decay of the long-lived ($t_{1/2} \approx 6$ h) isomer in ^{99m}Tc. Each transition is indicated with its respective total conversion coefficient, α_{TOT}. The $7/2^+ \to 9/2^+$ transition has a modest conversion coefficient ($\alpha_{\text{TOT}} = 0.119$) because the transition energy is relatively large and the multipolarity of the transition (M1) is low. By contrast, the hindered $1/2^- \to 9/2^+$ transition has a large conversion coefficient ($\alpha_{\text{TOT}} = 40.9$) due to the high multipolarity of the decay. The $1/2^- \to 7/2^+$ transition has an extremely large conversion coefficient ($\alpha_{\text{TOT}} = 1.4 \times 10^{10}$) due to its low energy and high multipolarity. Right: A cartoon of the internal conversion process where transitions occur by ejection of electrons from different electron shells, shown here schematically as clouds of K, L, M shell electrons.

$$\alpha_{ic} = \frac{T_{ic}}{T_\gamma} \tag{1.24}$$

where T_{ic} and T_γ are the electron emission and gamma-ray emission rates, respectively. Internal conversion can occur from different atomic shells. The probability of emission from specific shells generally depends on the multipolarity of the transition with a favouritism for innermost shells, e.g. K, L. Conversion coefficients are conventionally broken down into these different contributions where

$$\alpha_{\text{TOT}} = \alpha_K + \alpha_L + \alpha_M + \cdots \tag{1.25}$$

Internal conversion coefficients scale with Z^3 so internal conversion is most commonly associated with heavy nuclei. Internal conversion also depends strongly on transition energy, being enhanced for low transition energies. Moreover, internal conversion becomes more significant for high multipolarity transitions, e.g. M4 etc. Conversion coefficients can vary over many orders of magnitude even for transitions in the same nucleus; figure 1.16 shows an example of this focussing on the decay of the long-lived ($t_{1/2} \approx 6$ h) isomer in ^{99m}Tc—a decay exploited in medical imaging, particularly single photon emission computed tomography (SPECT).

A special case of internal conversion to note corresponds to internal pair transitions where an emitted photon converts to a positron–electron pair in the field of the emitting nucleus. Such a mechanism is only energetically possible above a transition energy of 1.022 MeV but, in general, only becomes significantly competitive to gamma-ray emission at higher energies, e.g. 2 MeV and above. Since relatively few transitions are associated with such high energies—the cases being generally limited to light nuclei—internal pair transitions are a somewhat unusual decay mode.

1.1.3.8 Fission

Fission is another possible mode of decay for heavy nuclei where the nucleus breaks up into two massive fragments, generally accompanied by emission of several excess neutrons. Some nuclei are sufficiently unstable with respect to fission that even their ground state may decay by this process—known as *spontaneous fission*. The probability of such a branch increases generally with Z; an example of a nucleus with a significant spontaneous fission branch is ^{252}Cf where the decay branch is around 3%. More generally, heavy nuclei may decay from excited states by fission. They can reach such high-lying fissioning states through other decay processes such as beta decay where the combined process is called *beta-delayed fission* or by a nuclear reaction which produces a nucleus in a high-lying state, e.g. *neutron-induced* fission.

The physics of fission is somewhat complex and aspects of it are still the subject of extensive experimental and theoretical study. Broadly speaking, the underlying physics relates to the quantum-mechanical phenomenon of tunnelling already noted as underpinning the physics of alpha decay (see figure 1.17). Even when the nucleus does not have enough energy to overcome the classical energy barrier (the strong nuclear force holding it together against the electrostatic repulsion of the protons), there is a non-zero probability for it to 'tunnel' through the barrier and split into two

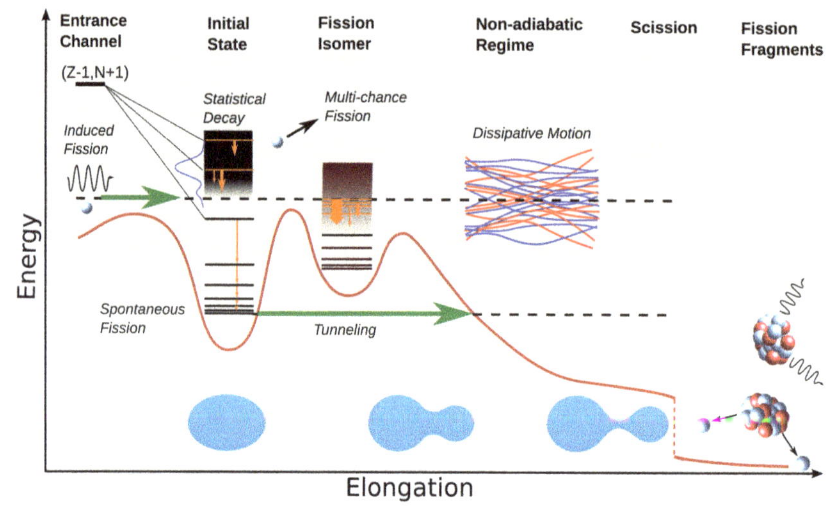

Figure 1.17. Schematic of some of the key features related to the phenomenon of fission. The red curve depicts (in a one-dimensional projection) the potential energy as a function of elongation; the ground state is the lowest minimum, and the shape-isomeric state is at the second minimum. From these states it is possible to tunnel through the potential barrier. Tunnelling is also relevant for neutron- or photon-induced fission when the resulting initial state lies below the fission barrier. If the initial state is excited above the fission barrier, it may undergo a complicated shape evolution crossing the barrier from above. Once the system finds itself beyond the barrier, it relatively quickly descends towards scission where it divides into two nascent fragments, which subsequently move apart under the influence of their mutual Coulomb repulsion, while gradually attaining their equilibrium shapes and becoming primary fragments. Primary fragments then de-excite by evaporating neutrons, emitting gamma-rays, and subsequently undergoing β decay. (Reproduced with permission from [19]. Copyright 2020 The Author(s). Published by IOP Publishing Ltd. CC-BY 4.0.)

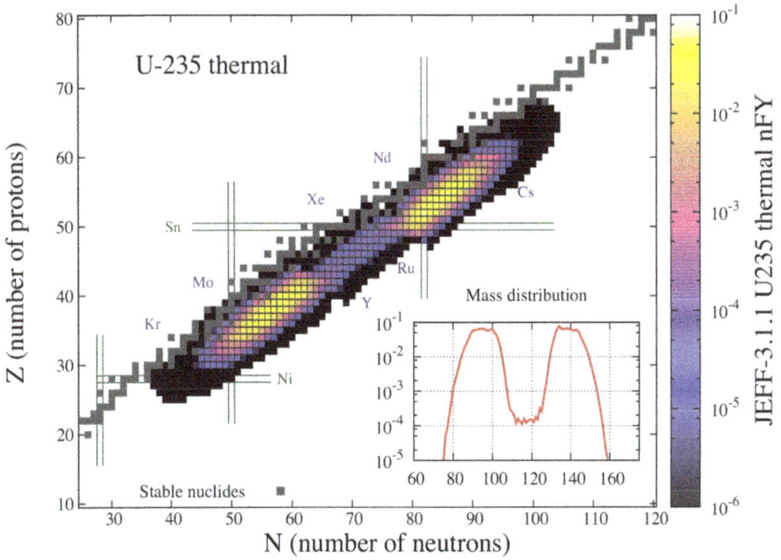

Figure 1.18. Fission fragment distribution from thermal neutron-induced fission of ^{235}U—the underlying process behind conventional fission reactors. Data are taken from the JEFF-3.1.1 neutron-induced fission yield file. (Reproduced with permission from [20]. Copyright 2017 Elsevier.)

or more smaller nuclei (fission fragments), releasing energy in the process. Most commonly, the fission fragments are asymmetric (see figure 1.18) meaning that there is one heavy and one light partner but the fission of exotic nuclei can sometimes lead to two more symmetric fission fragments. The neutron-rich fission fragments then typically decay by several successive β^- decays to reach the line of stability.

1.2 Nuclear reaction theory

Nuclear reactions are where two nuclei or a nucleus and a subatomic particle (such as an electron, photon, neutron, etc) collide, leading to a transformation of one or both of the nuclei involved. By transformation, we mean that the nuclei may end up changing their constituent number of protons and neutrons. Nuclei can also collide leading solely to a change in their initial direction—elastic scattering—and/or a change their direction and their internal excitation energy—inelastic scattering. The latter processes are not formally considered nuclear reactions but rather as two types of scattering process.

Nuclear reactions achieve the alchemist's dream of turning one element into another. Arguably, the study of nuclear reactions began with the work of Cockcroft and Walton in 1932 [21] who fired high energy protons at a sample of ^7Li turning some of the ^7Li nuclei into the unstable isotope ^8Be which promptly breaks up into alpha particles (^4He nuclei). This was the first time that scientists were able to 'split the atom' artificially and on demand.

Our understanding of nuclear reactions is underpinned by nuclear reaction theory. While this branch of theory is a distinct topic from nuclear structure discussed in the previous sections, there are, in fact, several ways in which the nuclear structure of the nuclei involved may influence reaction cross sections including: the *spin and parity* J^{π} of the ground state, the *shape* or *deformation* of the nucleus, the spectra of low-lying *discrete excited states*, and the *density of states function* $\rho(E, J, \pi)$ for states in the continuum region. The connections between nuclear structure and nuclear reaction theory are therefore somewhat profound and will be highlighted below as we give a brief overview of nuclear reaction theory of most relevance to the aims of this book. Although the mathematics of nuclear reactions will be explored, rigorous derivations will not be given and the reader wanting to approach the subject in more depth will find it useful to refer to textbooks covering nuclear reactions relevant to nuclear astrophysics, i.e. those underpinning nucleosynthesis in different stellar environments [22–24].

1.2.1 Definitions

Let us begin our discussion of nuclear reactions by focusing on nomenclature. A shorthand notation is commonly employed to denote different classes of nuclear reaction. This has the general form:

Target nucleus (projectile, ejectile(s)) **Residual nucleus**

e.g. ^{90}Zr(n,2n)^{89}Zr or ^{238}U(n,γ)^{239}U.

This form, once the target nucleus is obvious from the context, may be further reduced to

(projectile, ejectile(s))

e.g. (n,2n) or (n,γ).

Table 1.1 provides shorthand notation for a range of common nuclear reaction types involving incident neutrons. This general scheme is readily expanded to

Table 1.1. Descriptions of some common neutron-induced reactions alongside their shorthand labels.

Partial reaction	Description
(n,n) or (n,elastic)	A neutron is elastically scattered from the target nucleus
(n,n′) or (n,inelastic)	A neutron is inelastically scattered off the target nucleus leaving the nucleus in an excited state
(n,γ)	A neutron is captured by the target nucleus and one or more γ-rays are emitted
(n,2n)	One neutron enters the nucleus and two are emitted
(n,3n)	One neutron enters the nucleus and three are emitted
(n,f) or (n,fission)	The neutron causes the target nucleus to fission into two or more fragments
(n,α)	A neutron enters the nucleus and an α particle is emitted
(n,p)	A neutron enters the nucleus and a proton comes out
(n,np)	A neutron enters the nucleus and a neutron and a proton are emitted

reactions induced by other particles such as protons, e.g. (p,α), electrons, e.g. (e,e′p), or real photons, e.g. (γ,p) [25]. It should be noted that all of the reactions listed, with the exception of elastic scattering, may also lead to the emission of γ-rays from the residual excited nucleus.

1.2.2 Nuclear cross section and its behaviour as a function of energy

Let us continue our discussion of reactions by examining what is meant by *nuclear cross section*; a quantity which expresses the likelihood or probability that a given nuclear reaction or scattering process will take place. In a classical picture, we would treat an incoming particle such as a neutron as an effectively point-like particle colliding with a spherical nucleus of radius r. The likelihood of such an interaction would then be represented by the geometrical cross-sectional area of the nucleus, πr^2, that is 'seen' by the incoming particle. Characteristic values for the nuclear radius as a function of atomic mass, A, were given earlier in equation (1.4); they lie in the range $10^{-15} \rightarrow 10^{-14}$ m. It follows from this that the corresponding order of magnitude for the geometrical cross section of the nucleus is 10^{-28} m^2 or 10^{-24} cm^2. This led to the historical designation of the barn (b) as the unit for cross section[9] where $1\ b \equiv 10^{-28}$ m^2. While far from being an SI unit, the barn is an extremely useful yardstick and is used ubiquitously to define reaction probabilities[10].

The simple geometric estimate described above provides an order of magnitude for cross sections, but the detailed quantum-mechanical description of nuclear reactions is significantly more complex. Indeed, reaction cross sections can be significantly smaller or larger than the geometrical area of the nucleus. A good example of this would be the thermal neutron capture cross sections for some specific isotopes which exceed the simple geometrical estimate by more than five orders of magnitude (see figure 1.19). Such observations seem counter-intuitive, which demonstrates the limitations in adopting a semi-classical picture to think about the interactions of a complex, quantum-mechanical system such as the atomic nucleus.

Let us focus initially on discussing the characteristic behaviour of neutron-induced reactions as a function of energy. The total reaction cross section as a function of energy for neutrons incident on a range of different isotopes from ^{6}Li to ^{241}Am is shown in figure 1.20; the data are taken from the ENDF/B-VIII evaluated nuclear data library[11]. Let us examine the behaviour in the heaviest, actinide nuclei ^{235}U and ^{241}Am. Here, the total neutron cross section exhibits quite different trends in distinct energy regions. At very low or *thermal* energies, the cross section displays a smooth $1/E$ behaviour. In the energy range of a few electron volts (eV) to several kiloelectron volts (keV), the cross section displays a rapidly varying *resonant*

[9] The origin of this unit relates to a pun on the phrase as 'big as a barn door'.

[10] The millibarn (mb) is also frequently appropriate for reporting neutron-induced reaction cross sections.

[11] Around the world, several different collaborations of nuclear data physicists compile, maintain, and release libraries of *evaluated* (that is; extensively scrutinized and tested) nuclear data files for a large number of isotopes. The collaboration behind the ENDF/B series of libraries is mostly US-based, with the other major libraries being JEFF (European) and JENDL (Japanese).

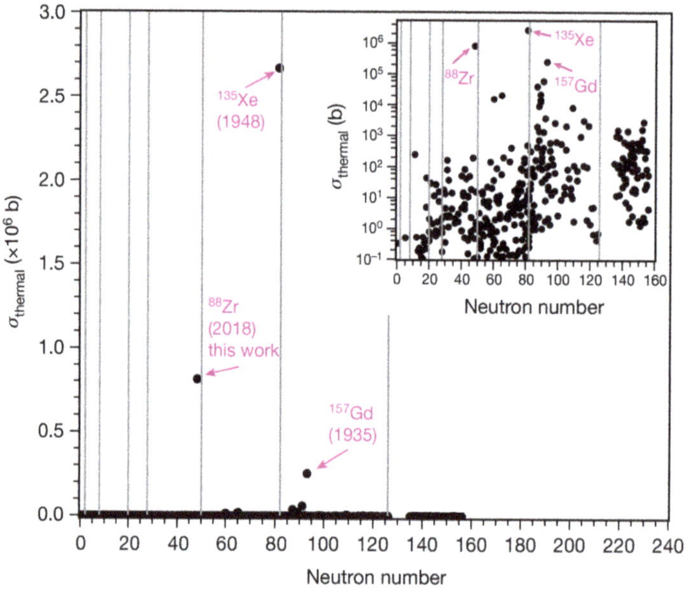

Figure 1.19. Thermal neutron capture cross sections as a function of neutron number on a linear plot and, inset, on a logarithmic plot. The position of the magic numbers for neutrons are marked with grey vertical lines. Examples of isotopes with very large neutron capture cross sections, i.e. ^{88}Zr, ^{135}Xe and ^{157}Gd are indicated. (Reproduced with permission from [26]. Copyright 2019 Springer Nature.)

structure where the energy available to the absorbed neutron coincides with discrete energy levels of the compound neutron-plus-nucleus system. This reaction mechanism persists into the so-called unresolved resonance region but here the levels of the compound nucleus system are either too finely spaced to show distinct structure, or are unable to be resolved experimentally. Finally, all fine details of nuclear structure become irrelevant and the cross section is again smooth. Total neutron cross sections for other nuclei shown in figure 1.20 exhibit broadly similar behaviour but qualitative differences reflect nuclear structure issues such as level density. The most different case is for ^{6}Li where the level density is very low and, hence, the role of resonances is minor.

Let us try to understand the features noted in figure 1.20 from the point of view of nuclear structure. As noted in section 1.1.2, the excited states of a nucleus are quantized. The consequence of this, at low excitation energies, is that the nucleus occupies certain discrete energy levels which have a narrow (energy) width, Γ. The existence of these discrete levels leads not only to the resonant behaviour of the cross section at low energies, but also the decreasing $1/E$ shape at very low energies.

Most neutron-induced reactions are *compound* nuclear reactions [28]. In these reactions, the incoming neutron is absorbed into the A-body target nucleus to form an $(A + 1)$-body compound nucleus. The energy from the absorbed neutron is shared amongst all the nucleons present leading to an excited compound nucleus state. The angular momentum of the incident neutron will also combine with that of

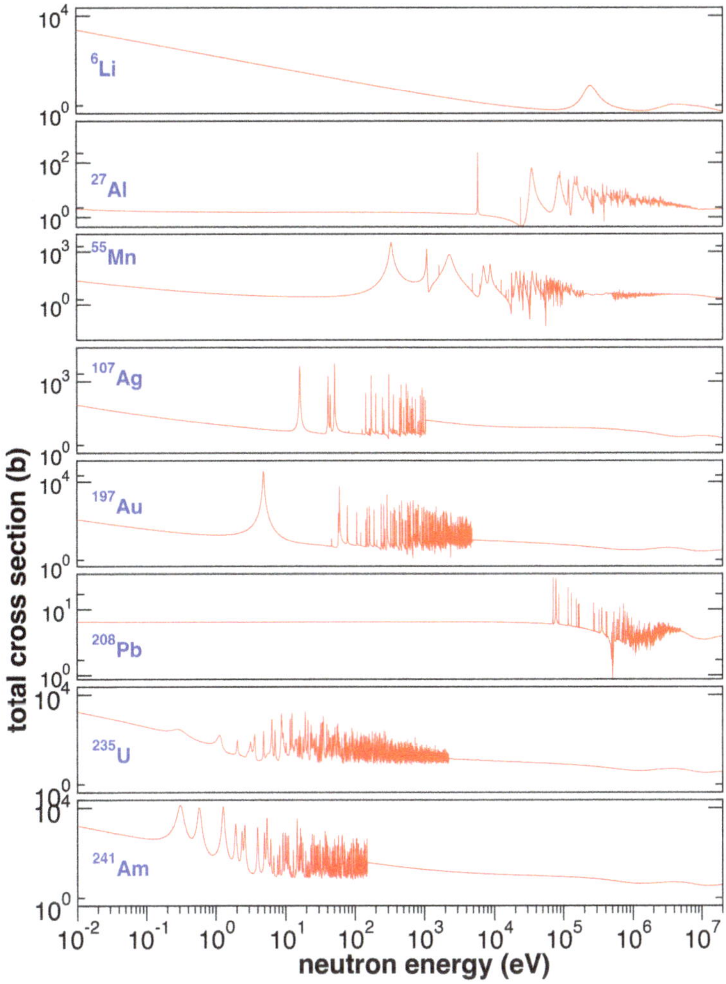

Figure 1.20. Total neutron cross sections as a function of incident neutron energy for a wide range of nuclei from ^{6}Li to ^{241}Am; data are taken from the ENDF/B-VIII library. (Reproduced with permission from [27]. Copyright 2010 Springer Nature.)

the target nucleus to potentially create a compound nucleus with a different spin/parity J^{π}.

When a neutron is captured by a target A-mass nucleus (in its ground state), the temporary $(A + 1)$-mass compound nucleus formed will be excited to the neutron separation energy S_n of the $(A + 1)$ nucleus plus any kinetic energy the neutron possessed. Therefore, if a neutron with very low kinetic energy is captured, the compound nucleus will have an excitation energy $\approx S_n$. S_n invariably coincides with the upper energy tail of a (negative relative energy) bound state and as such the cross section will be enhanced, as for other resonances, but decreases as incident energy increases (i.e. the observed $1/E$ behaviour) and the probability of overlapping with the bound state decreases.

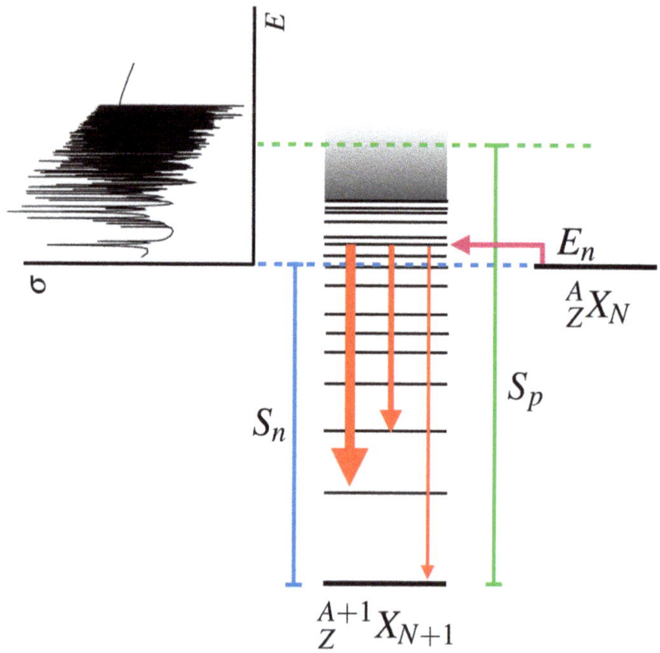

Figure 1.21. An illustration of the effect of discrete nuclear energy levels on the shape of the neutron reaction cross section. Here, neutron capture occurs on the nucleus ^{A}X for neutrons with energy E_n coinciding with some specific discrete level in the nucleus ^{A+1}X. The compound system, ^{A+1}X is shown schematically with the level density rapidly increasing. Being below the particle decay thresholds, e.g. $E_x < S_p$ as shown in the figure, the state populated in the neutron capture will gamma decay. The gamma decay is shown schematically here with the red arrows. The details of the decay branching will depend on gamma-ray selection rules (see section 1.1.3) and nuclear structure considerations.

Each quantized excited level of the $(A + 1)$ compound nucleus has a specific J^{π} assignment. In the resonance energy region the cross section is enhanced when the incident neutron energy (plus S_n) coincides with the energy of one of these discrete levels and the angular momenta of the target nucleus and neutron can combine in a vector sum to match the J^{π} of the state. This enhancement is responsible for the resonance structure. Figure 1.21 illustrates this coincidence between incident neutron energy and possible discrete excited states.

As the excitation energy increases, the spacings between these discrete levels become smaller and their energy widths Γ become wider, until eventually the levels overlap and form a continuum of states. The unresolved resonance region is that in which discrete states exist, but can no longer be distinguished experimentally, whilst the fast region is where the overlapping and broadening of the states has formed a true continuum. Moreover, with increasing excitation energy, it becomes energetically feasible for excited states to emit single or, potentially, groups of nucleons.

The total reaction cross section for an isotope gives the likelihood that any neutron-induced reaction will occur. The total cross section is constructed from the sum of the underlying *partial* cross sections which describe the likelihood of specific individual reactions occurring. Many different reactions will be possible, depending upon the energy available in the system. For neutrons, the elastic scattering process and radiative capture reaction are always energetically feasible, however the cross sections for other reactions will remain zero until the incident neutron has an energy above a certain threshold. Threshold reactions, e.g. (n,2n), (n,3n), etc, are, as expected, reactions which only occur above some threshold energy. This threshold energy will be equal to that required for emission of the outgoing particles. For example, the (n,2n) reaction, where the incident neutron is absorbed followed by two neutrons being ejected, requires the incident neutron to provide enough energy to overcome the binding energy of two neutrons. Figure 1.22 illustrates how the total cross section, in this case for neutron-induced reactions on the heavy nucleus ^{238}U, can be broken down into a set of partial cross sections for different reaction channels. The typical shape for a threshold reaction cross section is illustrated in figure 1.23. As may be seen, the cross section is smoothly varying; it is zero below its threshold energy and peaks before slowly falling off. The fall off following the peak is due to other threshold reactions 'switching on', i.e. becoming energetically feasible as the excitation energy of the compound nucleus increases and so the total cross section strength becomes shared between a greater number of channels.

Nuclei may be excited into isomeric states through neutron-induced reactions, e.g. inelastic scatter, or an isotope may be transmuted via, e.g. (n,2n), into an excited

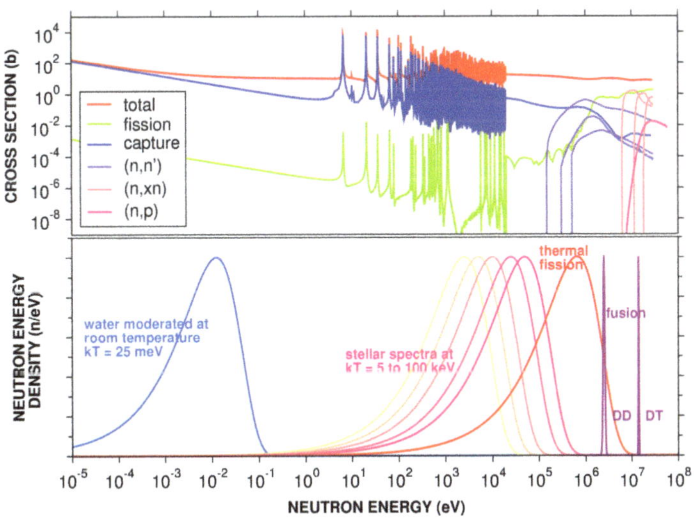

Figure 1.22. Top panel: Neutron-induced reaction cross sections for ^{238}U as a function of neutron kinetic energy. Bottom panel: Examples of characteristic neutron energy distributions found in different astrophysical environments and technological applications. Neutron spectra for stellar environments at a range of temperatures from $kT = 5$ to 100 keV are shown. Neutron spectra from DD and DT fusion are shown in purple (see chapter 2) while the neutron energy distribution from a typical fission reactor is shown in red. A water moderated thermal neutron spectrum ($kT = 25$ meV) is shown in blue. All distributions have been normalized to their maximum value for ease of presentation. (Reproduced with permission from [29]. Copyright 2018 Elsevier.)

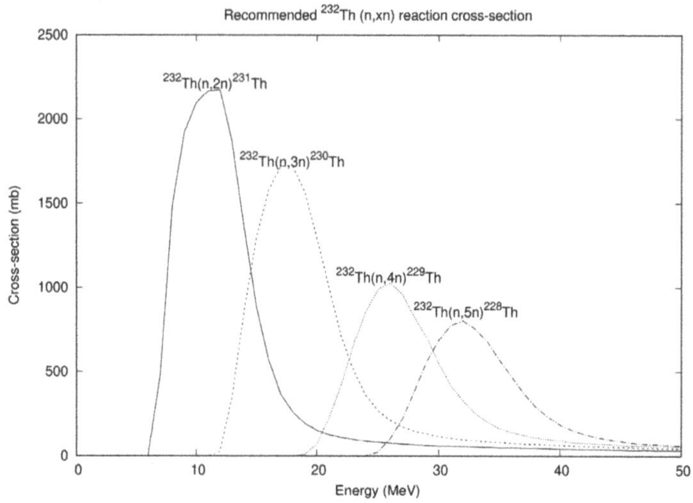

Figure 1.23. Example behaviour of threshold reaction cross sections. The cross section data shown are for (n,xn) reactions on ^{232}Th extracted using TALYS 1.9. (Reproduced with permission from [30]. Copyright 2019 Elsevier.)

product isotope which then decays, typically by γ decay, with the γ cascade 'stalling' as it passes through an isomeric state. Whilst an isotope is in an isomeric state, its reaction cross sections will be different relative to those of its ground state. If the isomer is likely to be populated in significant quantities and exist for a relatively long time in a nuclear system being studied, its cross sections should be included in the nuclear data library used to model that system.

1.2.3 Direct and pre-equilibrium reactions

Another type of reaction is the *direct* reaction, where an incident particle only interacts with one or two of the nucleons present in the target nucleus. This reaction is less common for incident neutrons than it is for charged particles. Compound nucleus reactions generally occur over timescales of $\approx 10^{-15}$ s, whereas direct reactions have a far shorter interaction period of $\approx 10^{-21}$ s.

Pre-equilibrium reactions[12] are an intermediate point between direct and compound. In these reactions, the incident neutron interacts and shares its energy with a significant fraction of the other nucleons present, but the system does not become equilibrated as in the compound case.

As well as being able to differentiate between these reaction types using timescales, each also emits outgoing (or secondary) particles with differing distributions of both energy and angle. Figure 1.24 illustrates the differences in secondary particle energy distributions.

[12] Also referred to as pre-compound or multi-step processes.

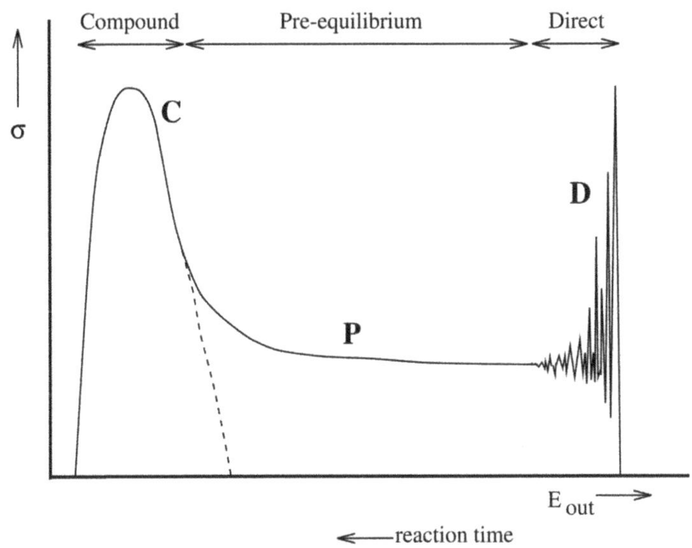

Figure 1.24. An illustration of the differences in the energy of particles emitted in compound nuclear, direct, and pre-equilibrium reactions. (Reproduced with permission from [31]. Copyright 2023 Springer Nature.)

In terms of angular distributions, outgoing particles will be emitted almost isotropically (in the centre-of-mass frame) for compound reactions, but will be emitted more favourably at forward angles for direct and pre-equilibrium reactions. As discussed later in this chapter, the angular distributions of outgoing particles in direct reactions can give information about the underlying nuclear structure of the target nucleus.

1.2.4 Bound state and continuum wave functions

Wave functions are at the heart of quantum mechanics and describe the quantum state of a system. In nuclear physics, wave functions are used to understand the properties of nucleons within a nucleus, as well as interactions between nuclei. These wave functions are categorized as bound state and continuum wave functions, depending on the energy of the system relative to the potential in which the particles reside. Here, we will explore these concepts, focusing on their role in nuclear physics. The interested reader will find more on quantum mechanics in nuclear structure physics elsewhere in this IOP Book Series [32, 33].

1.2.4.1 Bound state wave functions

A bound state is a quantum state where the particle remains confined to a potential well due to an attractive force, such as the nuclear potential. The energy of a bound state is negative, indicating that the system has less energy than a free particle at rest, which prevents it from escaping the potential well. Bound states are characterized by discrete energy levels, and the wave functions associated with them are localized within the nuclear potential well.

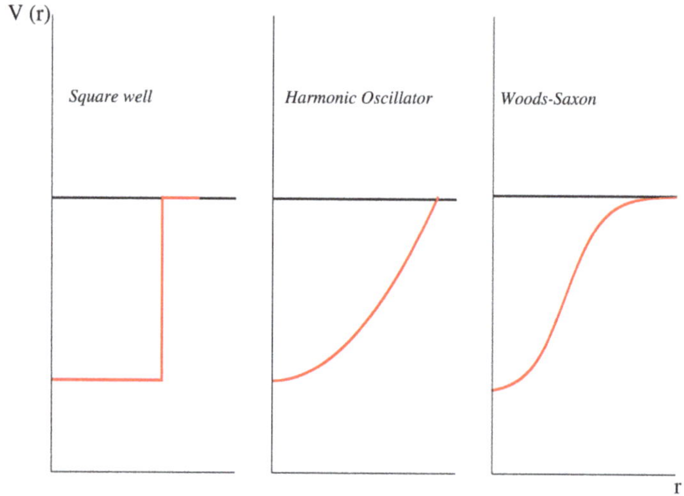

Figure 1.25. Schematic representation of the square well, harmonic oscillator and Woods–Saxon potentials.

Bound state wave functions are solutions to the time-independent Schrödinger equation given by

$$\hat{H}\psi = E\psi, \tag{1.26}$$

where $\hat{H}$ is the Hamiltonian operator, representing the total energy of the system, ψ is the wave function and E is the energy eigenvalue, corresponding to the energy of the bound state. The Hamiltonian for a nucleus typically includes kinetic energy terms for each nucleon and a potential energy term describing the interaction between nucleons. The potential energy term is complex and is often modeled using an effective (or 'mean-field') nuclear potential, such as a square well, harmonic oscillator or Woods–Saxon potential (see figure 1.25). Simpler options such as a finite square well or a harmonic oscillator are useful for illustration of the shell structure in nuclei, but the most commonly used for practical nuclear structure and reaction calculations is the Woods–Saxon potential which usually consists of central and spin–orbit components.

When dealing with heavier exotic nuclei (with a large neutron-to-proton ratio), additional potential terms may be needed, such as deformation and Coulomb potentials. The Coulomb potential, which is repulsive for charged particles, extends outside the nucleus. Instead of adding it directly to the potential in the Schrödinger equation, its effect can be included by modifying the wavefunction in the outer region (asymptotic region).

For a spherically symmetric potential $V(r)$, the time-independent Schrödinger equation is given by

$$\left[-\frac{\hbar^2}{2\mu}\frac{\mathrm{d}^2}{\mathrm{d}r^2} + V(r)\right]\psi(r) = E\psi(r), \tag{1.27}$$

where $\hbar$ is the reduced Planck's constant, $\psi(r)$ is the radial wave function, $V(r)$ is the potential energy as a function of radius, E is the energy eigenvalue and μ is the reduced mass of the system, given by

$$\mu = \frac{m \cdot M}{m + M}, \tag{1.28}$$

where m and M are the masses of the bound particle and the sum of the other nucleons present in the nucleus respectively. Using the reduced mass instead of individual masses, allows the two-body problem to be solved as a one-body problem. Solutions to equation (1.27) corresponding to bound states are where $E < 0$.

The bound state wave function decays exponentially outside the range of the potential ($r \to \infty$), indicating that the probability of finding a nucleon at large distances from the nucleus is negligible and is given by

$$\psi(r) \propto e^{-kr}, \quad k = \sqrt{\frac{-2\mu E}{\hbar^2}}. \tag{1.29}$$

As can be seen, k in equation (1.29), which may be thought of as the wavenumber of the state, will be an imaginary quantity. ψ exhibits discrete energy levels, corresponding to the quantization of energy in the potential well. Bound state wave functions are normalized, meaning the integral of the square of the wave function over all space is equal to one. This reflects the probability of finding the nucleon within the nucleus.

The deuteron—a bound state of a neutron and a proton—is a classic textbook example. Its binding energy is approximately 2.22 MeV. The nuclear potential can be modeled as a Yukawa potential given by

$$V(r) = V_0 e^{-r/a}, \tag{1.30}$$

where V_0 and a are constants representing the depth and range of the potential, respectively. The wave function of the deuteron is localized within the range of the potential.

1.2.4.2 Continuum wave functions

In contrast to a bound wave function, a continuum wave function represents a state where the particle has enough energy to escape the potential well ($E > 0$). These states are not bound and are associated with scattering processes.

The same time-independent Schrödinger equation applies, but for $E > 0$, the solutions are oscillatory

$$\psi(r) \propto e^{ikr} + f(r)e^{-ikr}, \tag{1.31}$$

where $k = \sqrt{\frac{2\mu E}{\hbar^2}}$ and $f(r)$ is the *scattering amplitude.*

The wave function oscillates as ($r \to \infty$), reflecting the free-particle nature. A number of different mathematical representations may be used for the wave function, such as

$$\psi(r) \propto \sin(kr + \delta(k)), \tag{1.32}$$

where $\delta(k)$ is the *phase shift* due to the potential. The phase shift is related to the cross section and corresponds to a continuous spectrum of energies.

1.2.4.3 Non-normalization

Continuum wave functions are not normalizable in the traditional sense, as they extend to infinity. However, they can be normalized using a delta function normalization. Consider a neutron with kinetic energy E scattering off a nucleus as a classic example. The potential (in this case, the nuclear optical potential, described below in section 1.2.7), may include real and imaginary components,

$$V(r) = U(r) + iW(r), \tag{1.33}$$

where W accounts for the probability of non-elastic processes (e.g. compound nuclear reactions) occurring. The continuum wave function describes the likelihood of the neutron scattering elastically or being absorbed.

A practical use case of these concepts is in the study of resonances in nuclear reactions. Resonances occur when a continuum state temporarily behaves like a bound state, leading to enhanced scattering cross sections. For example, in ^{13}C (^{12}C + n), certain neutron energies correspond to resonant states in ^{13}C. These are quasi-bound states with positive energy but confined for a short time due to the potential barrier.

The interplay of bound state and continuum wave functions is crucial for understanding nuclear structure and reactions. Bound states describe stable configurations, while continuum states govern scattering and decay processes. Together, they provide a comprehensive framework for exploring phenomena such as nuclear binding, resonances, and reactions in nuclear physics.

1.2.5 Recipe to deal with continuum discretization

Continuum wave functions are functions of a continuous variable, i.e. the energy of continuum along with discrete spatial variables. There are only a few methods available on the market to deal with continuum wave functions [34]. The two methods used so far are the bin method [35] and the pseudostate method [36]. In each method, the continuum wave function is approximated by a finite set of square-integrable functions.

1.2.5.1 Bin method

In the bin method, the continuum is discretized into a finite number of bins or sets of energy intervals; the internal states within each bin are represented by a single wave function, which is constructed by superposition of true continuum wave functions within that bin. It is further categorized into two types:

- The **mid-point method** [35, 37] consists of taking scattering state

$$\tilde{\psi}_i(r) = \sqrt{\Delta}\ \psi(r, \bar{E}_i), \quad E_i > 0 \tag{1.34}$$

for a discrete set of scattering energies, where $\bar{E}_i = (E_i + E_{i-1})/2$, with Δ as the common energy interval or width of the bin. In the mid-point method, continuum channels are represented by the channel at the mid-point of the bin. The resulting set of wavefunctions $\tilde{\psi}_{ij}(r)$ satisfies the following orthogonality condition

$$\int \tilde{\psi}_i(r)\tilde{\psi}_j(r)\mathrm{d}\vec{r} = \Delta\ \delta_{ij}\delta\big(\bar{E}_i - \bar{E}_j\big), \tag{1.35}$$

that depends on the bin width Δ.

- The **average method** [37–40], where scattering wave functions $\tilde{\psi}_i(r)$ are averaged over E_i to make them square integrable and is given by

$$\tilde{\psi}_i(r) = \frac{1}{\sqrt{\Delta}} \int_{E_{i-1}}^{E_i} \psi(r, E)\mathrm{d}E. \tag{1.36}$$

In the average method, the continuum channels within each bin are averaged into a single channel. The resulting set of wavefunctions $\tilde{\psi}_{ij}(r)$ satisfies the following orthogonality condition

$$\int \tilde{\psi}_i(r)\tilde{\psi}_j(r)\mathrm{d}\vec{r} = \delta_{ij}\delta_{E_i E_j}, \tag{1.37}$$

which is free from any explicit dependence on Δ.

1.2.5.2 Pseudostate method

In the **pseudostate (PS) method** [36, 39, 41], PSs are used, which are just the eigenstates of the internal Hamiltonian on some convenient square-integrable basis. A variety of PS bases are available in the literature for two-body continuum discretization [42–44] and also for three-body continuum [45, 46]. These PSs decay to zero at large distances.

The PS method excels when dealing with narrow resonances, and allows for a detailed representation of resonance profiles using a relatively small basis, facilitated by convolution techniques. Conversely, the binning method necessitates a finer discretization to accurately capture the resonance region within the relative energy spectrum. While both methods should yield consistent reaction observables when fully converged, practical considerations often determine the preferred approach.

1.2.6 Calculating cross sections

Having had an excursion into some of the ways in which wave functions are handled in nuclear reaction theory, we return to the topic of calculating cross sections. Again, we will mostly limit ourselves to the case of incident neutrons.

An incident flux of neutrons can be represented by a quantum-mechanical wavefunction, e.g. ψ_i. In order for an interaction with a target nucleus to occur, there must be a potential present (due to the strong nuclear force), e.g. V. If a reaction occurs, the outgoing neutrons will be represented by a new wavefunction, e.g. ψ_f. In the direct reaction case, ψ_f may represent a completely different outgoing particle, e.g. in the (p,n) reaction ψ_i would represent an incident proton and ψ_f would be an outgoing neutron.

For compound nuclear reactions, ψ_f will represent neutrons that have either not interacted with or have elastically scattered from V. By examining the changes in the phase of the wavefunction between ψ_i and ψ_f, i.e. the *phase shift* introduced earlier in equation (1.32) and which will be returned to later in equation (1.42), the elastic scattering cross section can be calculated. In addition, the amount of flux absorbed into the compound nucleus and hence the total of the non-elastic partial cross sections can also be determined.

1.2.7 The optical model

Different theoretical models and tools are used for different reaction types and energy regimes. For example, direct reactions are usually treated using the distorted wave Born approximation (DWBA) and pre-equilibrium reactions by *exciton* models [47]. Here, we restrict ourselves to looking in more detail at compound nuclear reactions.

We can write the cross section for a reaction which proceeds through a compound nucleus stage as

$$\sigma_{\alpha\beta} = \sigma_{\text{CN}} G_\beta, \tag{1.38}$$

where α and β are the entrance and exit *channels*, e.g. a single neutron in and two neutrons out, respectively, for the (n,2n) reaction. σ_{CN} is the cross section for compound nucleus formation and G_β is the branching ratio, or probability, for the compound nucleus to decay into exit channel β.

Although it is not the only method available, the *optical model* [48, 49] is the most widely used when calculating compound nuclear reaction cross sections (σ_{CN} in equation (1.38)) at energies above the resonance region. In reality, when an incident neutron approaches a target nucleus, it will be influenced by the potentials of each of the individual nucleons present. Rather than treat this complicated situation exactly, the optical model replaces these individual many-body potentials with a single *mean-field* potential which is constructed in such a way as to include all of the necessary interactions which may occur.

The optical model potential (OMP) contains a large number of parameters, which are generally determined for an isotope by conducting neutron scattering experiments. The Woods–Saxon potential (plotted in figure 1.25) is the most commonly used OMP for modelling neutron-induced reactions. The Woods–Saxon potential includes terms to account for elastic scattering, absorption, and spin–orbit effects (in addition to a Coloumb term for charged incident particles).

For simplicity, here we will simply write a generic OMP as

$$V(r) = U(r) + iW(r), \tag{1.39}$$

where U is the real part of the potential responsible for elastic scattering and W is an imaginary part responsible for absorbing neutrons into the compound nucleus.

In order to calculate the changes to the wavefunction as a result of the interaction with the potential present, the (time-independent) Schrödinger equation must be solved

$$\hat{H}\psi = E\psi, \tag{1.40}$$

where $\hat{H}$ is the *Hamiltonian* operator for the system. Applying it, as in equation (1.40), to ψ, will give the total energy of the system. $\hat{H}$ is composed of terms related to the kinetic energy T (and angular momentum) of the incident particle and the nuclear potential it experiences,

$$\hat{H} = \hat{T} + V. \tag{1.41}$$

In the optical model, the Schrödinger equation is solved in the region close to the target nucleus (the *interior* region), where the potential is present, and further away where the potential (with the exception of the Coulomb potential for charged particles) is taken to be zero (the *exterior* region). The solutions for ψ in each region must match smoothly at the boundary between the two regions. This matching is illustrated in figure 1.26 for a one-dimensional wavefunction.

In general, ψ is decomposed into a set of *partial waves*. The set of partial waves forms a set of orthonormal basis states from which ψ is constructed, with each partial wave representing incident particles with a specific (orbital) angular momentum value l. The angular momentum of the neutron–nucleus interaction is due to the linear momentum of the incident neutron and the perpendicular distance (the *impact parameter*) at which it passes or interacts with the target nucleus. Figure 1.27 highlights the generation of different angular momentum values depending on the perpendicular distance between the neutron and target nucleus.

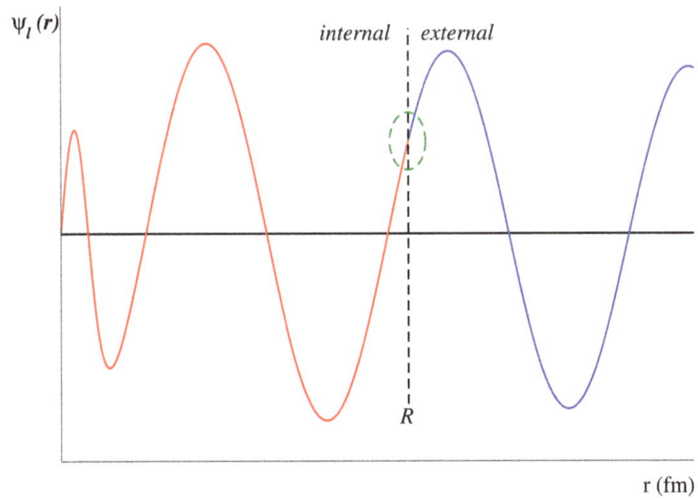

Figure 1.26. Schematic representation of the radial wave function, separating the inner (red) and outer part (blue).

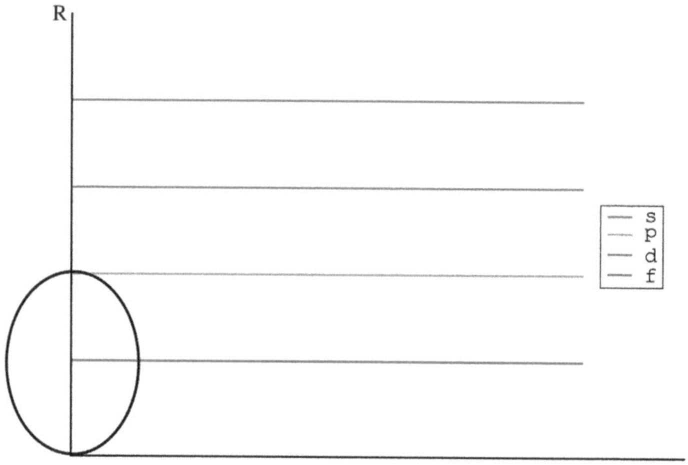

Figure 1.27. The generation of different angular momentum values depending on the perpendicular distance (R) between the neutron and target nucleus.

The lower values of angular momentum are generally labelled using traditional spectroscopic notation, i.e. s, p, d or f rather than $l = 0, 1, 2$ or 3, respectively. For incident energies below ≈20 MeV, only the first few partial waves need be considered and for lower energies, i.e. those in the thermal and resonance regions, *s*-wave reactions typically dominate.

The incident wavefunction ψ_i is modified by V to give ψ_f. As described previously, ψ_f can be used to determine the magnitude of elastic scattering that has occurred (through the phase shift) and hence the elastic cross section. ψ_f will also show the amount of flux absorbed into the compound nucleus. This can be converted to a total non-elastic cross section which will be equal to the sum of all of the other possible reactions.

The phase shift δ_l (for a partial wave l) can be used to create the *S-matrix element* S_l by

$$S_l = e^{2i\delta_l}. \tag{1.42}$$

S_l is a complex quantity. For a real potential ($W = 0$), where only elastic scatter is possible, $|S_l| = 1$.

S_l can be used to calculate the elastic scattering cross section via

$$\sigma_{\text{elas}} = \frac{\pi}{k^2}\sum_l (2l + 1)|1 - S_l|^2, \tag{1.43}$$

the cross section for non-elastic processes/reactions by

$$\sigma_{\text{reac}} = \frac{\pi}{k^2}\sum_l (2l + 1)(1 - |S_l|^2), \tag{1.44}$$

and the total cross section by

$$\sigma_{\text{tot}} = \frac{2\pi}{k^2}\sum_l (2l + 1)(1 - \text{Re}(S_l)). \tag{1.45}$$

In each of equations (1.43), (1.44) and (1.45), the sum is over the range of all incident partial waves. σ_{reac} can be taken as equivalent to the cross section for compound nucleus formation σ_{CN}.

1.2.8 The Hauser–Feshbach method

With the compound nucleus formation cross section σ_{CN} known from, for example, the optical model, a new method is then required to calculate the probability of decay (G_β in equation (1.38)) of the compound nucleus into the range of possible exit channels. The Hauser–Feshbach [50] method is typically used to do this.

Depending upon the energy and angular momentum of the incident neutron, the compound nucleus may have a range of possible excitation energies and spins. The *density of states* $\rho(E, J, \pi)$, an isotope-specific energy E, spin J, and parity π dependent function, will be used to determine the distribution of these excited states. A number of empirical and theoretical density of states functions are available for selection, with differing regimes of applicability and complexity for each. For the purpose of the calculation, the distribution of states is separated into different *bins*, each with a different set of E, J and π values.

The Hauser–Feshbach [50] method takes each possible bin and works out the quantum-mechanical overlap between that and a possible exit channel. In other words, it determines how likely the process is and therefore what proportion of reactions will lead to the emission of those particles. The *T-matrix element*, calculated using a suitable OMP for each outgoing channel, spin and energy (in a sort of reversal of calculating the cross section for compound nucleus formation), is the dominant quantity governing the magnitude of branching ratios. The T-matrix element T_l is related to S_l by

$$S_l = 1 + 2iT_l. \tag{1.46}$$

The branching ratio, in turn, is combined with the compound nucleus formation cross section σ_{CN} to give the cross section for a particular reaction channel. This process is repeated for each possible bin and exit channel. Separate Hauser Feshbach decays will also be needed for each different compound nucleus populated at the different steps of a decay chain.

As the excitation energies of the populated compound nuclei become low enough, the only possible decays will be via γ-ray emission. If the energies of the low-lying excited states of a nucleus are known, these γ decays can be calculated using more simplistic transition rules. Figures 1.28 and 1.29 illustrate possible decay paths from a compound nucleus above and below the various particle separation energies, respectively.

1.2.9 *R*-matrix

For reactions occurring in the resonance region or for low-mass nuclei, the optical model approach of utilizing a mean-field potential becomes less valid. This is due to individual excited states having more significant effects on the reaction process.

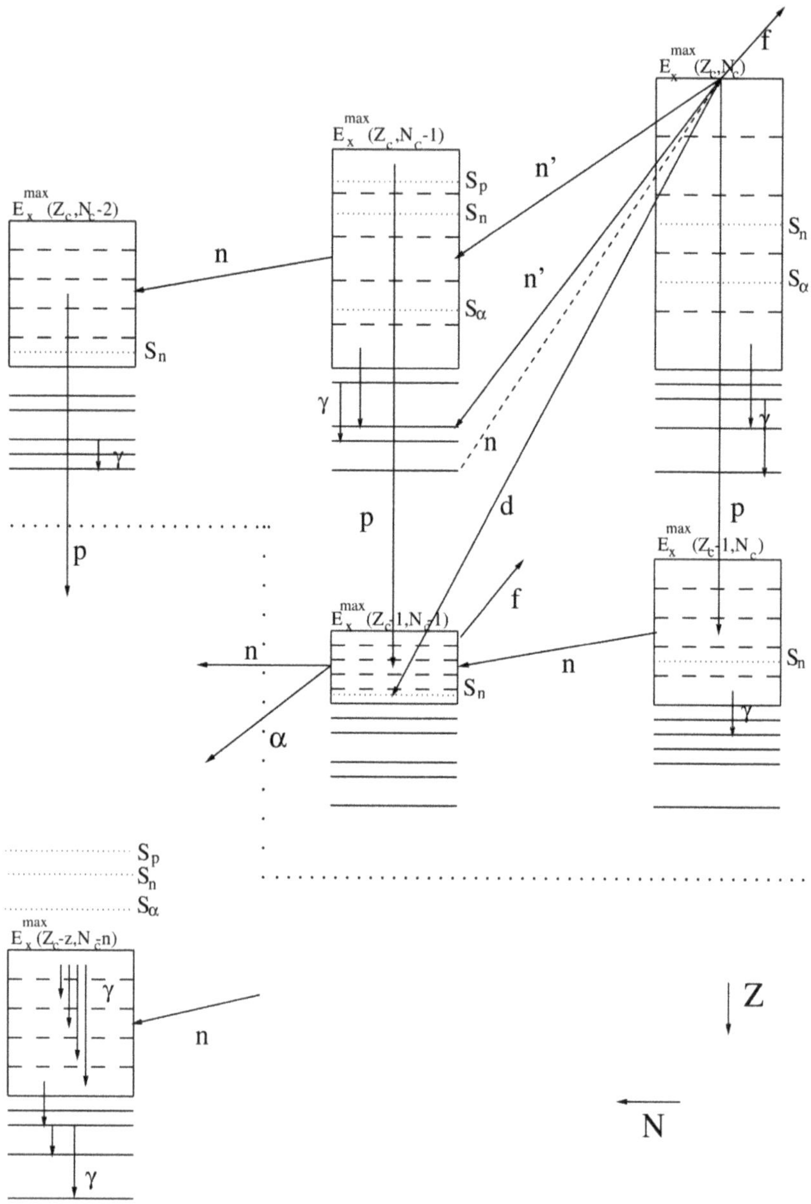

Figure 1.28. An illustration of the decay of a compound nucleus. The figure is in two parts. The lower left portion considers the case of neutron capture into a system below the particle decay thresholds; this scenario is shown more clearly in figure 1.29. The remainder of the figure considers the case where a compound nucleus is produced well above the particle decay thresholds. Followed from the very top right, the figure shows different potential decay pathways; see [31] for more discussion of the details in this busy figure. (Reproduced with permission from [31]. Copyright 2023 Springer Nature.)

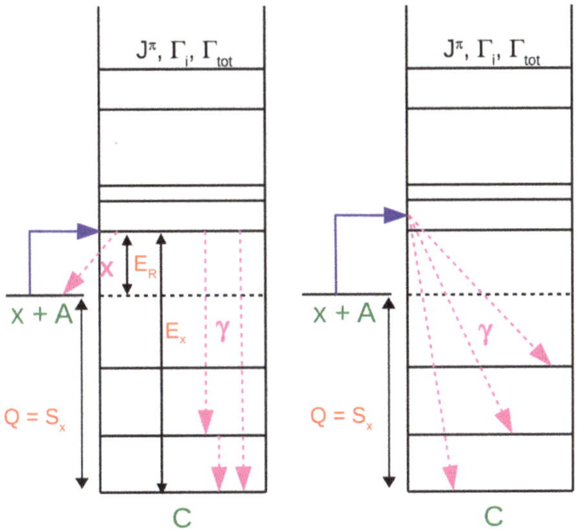

Figure 1.29. An illustration of the decay of a compound nucleus below the various particle separation energies. The decay proceeds entirely by gamma-ray emission (with internal conversion in competition). Left: Proton (or neutron) capture into a resonance state. Right: Proton (or neutron) *direct* capture, not through a resonance state. (Reproduced from [51]. CC-BY-4.0.)

Figure 1.30 illustrates the enhancement of the cross section at particular excited energies corresponding to resonances. In the region outside of the effect of the nuclear potential, $r > R$, the incident neutron can be modelled using a plane wave with an amplitude and wavelength determined by its kinetic energy. Inside the nucleus's potential, $r < R$, the neutron's wavefunction will be constrained by the shape and strength of the potential. At kinetic energies where the amplitude of the neutron's wavefunction will be significantly different in the two regions, the probability (and hence cross section) for it to penetrate into the target nucleus (and hence undergo a reaction) will be reduced. However, for energies where the amplitude is well matched in the two regions, the probability of the neutron penetrating into the nucleus will be large and the cross section will be enhanced and thus display a resonance.

The *R*-matrix method [53, 54] may be used to perform calculations of the cross section in the resonance region. In this method, the interior wavefunction is constructed from a set of orthonormal basis states. These states are constructed based on a number of boundary conditions where the interior and exterior regions of the calculation meet, as illustrated in figure 1.31.

The *R*-matrix can be used purely predictively, but it is far more commonly used as an extremely sophisticated data fitting tool. Measured resonance data for a target isotope are supplied as inputs and an *R*-matrix code, e.g. REFIT or SAMMY, iteratively adjusts the forms of the interior basis states and other parameters used to reproduce the measured data. The interior basis states are usually initially constructed using the properties of known excited levels in the compound nucleus. The *R*-matrix code will

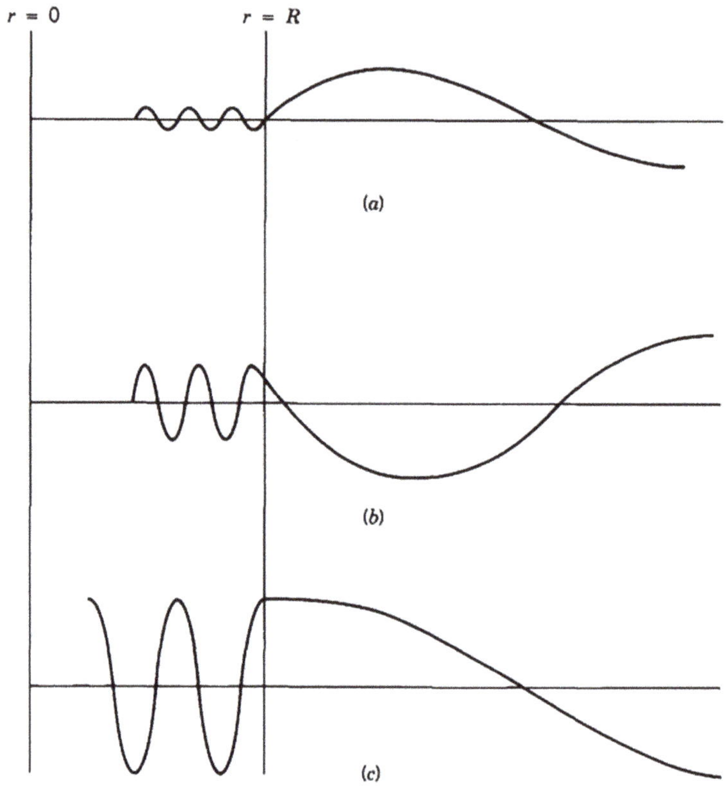

Figure 1.30. Illustration of the enhancement of the penetration of a neutron's wavefunction into the interior region when its amplitude matches that of the internal wavefunction. (Reproduced with permission from [52]. Copyright 1988 Wiley and Sons.)

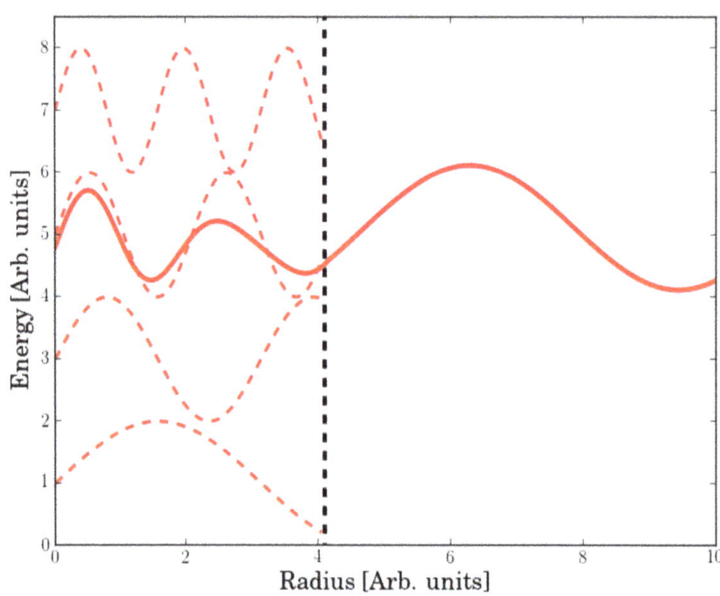

Figure 1.31. Illustration of the creation of the interior wavefunction from a set of orthonormal basis states.

then perform adjustments of the parameters of these states, within allowed bounds, to better fit the data. The adjusted parameters are then used to calculate additional unmeasured nuclear data quantities of interest for that particular target isotope.

The measured data supplied are typically neutron transmission and neutron capture measurements. In a transmission measurement, the neutron flux through a thin target material is measured and then the flux is measured with the material absent; the ratio of the results in each case is then taken as a function of energy. In a neutron capture measurement the target material is bombarded with neutrons and any emitted γ-rays are measured. In each case, a time-of-flight diagnostic may be used to determine the energies of neutrons either transmitted or absorbed (see later discussion of the nToF time-of-flight facility at CERN in section 5.2.1).

Cross sections for the resonance region are generally represented by *resonance parameters* rather than a series of cross section versus energy points. These parameters include the energy of the resonance E_0, its spin J (i.e. the total angular momentum of the compound nucleus formed), the orbital angular momentum l of the incoming neutron, and the partial widths Γ_n, Γ_γ for each possible reaction. By assuming a particular shape for the resonances, the cross section for each may be constructed on an arbitrarily fine energy grid. A *Breit–Wigner* form is usually adopted for the shape of the cross section near an isolated resonance:

$$\sigma \approx \frac{\Gamma_a \Gamma_b}{(E - E_r) + \Gamma_{\text{tot}}^2/4}, \tag{1.47}$$

where Γ_a, Γ_b, and Γ_{tot} are the widths for the projectile plus target nucleus, ejectile plus residual nucleus, and the total width, respectively.

An isolated resonance or a capture resonance have the general Breit–Wigner form, as illustrated in figure 1.32, but resonances for elastic scatter, fission and other reactions may be somewhat distorted due to scattering from the background potential (*shape elastic* or *potential* scattering) or interference effects from neighbouring resonant states.

The R-matrix is typically used at all energies for very light isotopes with a low density of states and at low energies for heavier isotopes where individual excited states may still be resolved separately.

1.2.10 DWBA method

For direct reactions, the model most commonly used is the DWBA. DWBA analysis is typically not too important for neutron-induced reactions as these generally proceed through an intermediate compound nucleus. However, incident neutrons may still take part in direct reactions and the DWBA method is extremely useful for determining the structure of target nuclei through charged particle reactions. A knowledge of nuclear structure is obviously important for correctly interpreting and predicting other reaction types.

In the DWBA approach, the cross section $\sigma_{\alpha\beta}$ for a reaction for an incident channel α leading to an exit channel β can be written as

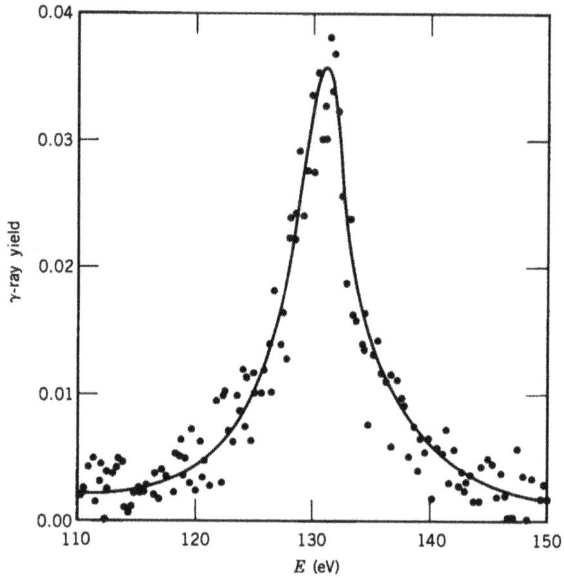

Figure 1.32. Example neutron capture cross section fitted through an experimental dataset showing a typical Breit–Wigner form resonance in ^{59}Co. (Reproduced with permission from [52]. Copyright 1988 Wiley and Sons.)

$$\sigma_{\alpha\beta} = B \cdot T_{\alpha\beta}^2, \tag{1.48}$$

where $T_{\alpha\beta}$ gives the overlap between the wavefunctions for the incident and final channels and the potential which mediates the transition. $T_{\alpha\beta}$ can be written as

$$T_{\alpha\beta} = \int V_{\alpha\beta}\psi_\alpha \psi_\beta^* \mathrm{d}R, \tag{1.49}$$

where R represents the spatial dimensions of the problem, or in Dirac's bra–ket notation as

$$T_{\alpha\beta} = \left\langle \psi_\beta \middle| \alpha\beta \middle| \psi_\alpha \right\rangle. \tag{1.50}$$

The potential $V_{\alpha\beta}$ used to calculate $T_{\alpha\beta}$ is different from the OMPs described earlier. $V_{\alpha\beta}$ is a short-ranged potential which will act to, for example, exchange the particle in the incident channel with another particle in the exit channel. OMPs are, however, still used in DWBA calculations. ψ_α and ψ_β are *distorted* wavefunctions calculated for the respective particle in the presence of an OMP. Calculating distorted waves takes into account any competing compound nuclear reactions which may otherwise interfere with the calculation of the direct cross section.

The short timescale of direct reactions, combined with the incident particle only interacting with one, or possibly two nucleons in the target nucleus, means that the (orbital) angular momentum, l, of the particle removed or scattered from the nucleus can be determined via an analysis of the angular distribution of the outgoing

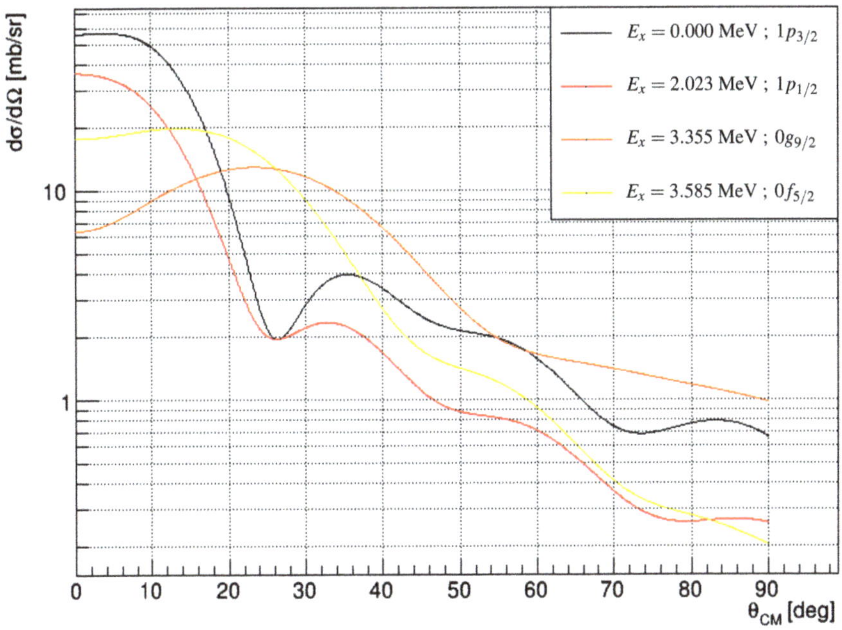

Figure 1.33. Angular distribution of protons from the ^{48}Ca(d,p) reaction as a function of their angle in the centre of mass. Four curves are shown for transfer to different excited states labelled by the excitation energy of the state and associated shell model state.

particle; an example of such angular distributions is shown in figure 1.33. By observing the decay of the residual nucleus by e.g. gamma decay, the energy level that the nucleon was removed from may also be determined. Systematically studying the target nucleus using a range of incident energies and the above techniques means information about the structure of the nucleus, in terms of the nuclear states populated, can be determined.

B in equation (1.48) is referred to as a *spectroscopic factor*. B is essentially a scaling factor which can be determined by comparing the measured versus calculated DWBA cross section. B is related to how 'pure' the nuclear state is for the nucleon (in the target nucleus) involved in the reaction, i.e. how close the state is to that expected from an extreme independent-particle model (see discussion of such models in [3], another volume in this series). When experimental data are unavailable, B may be estimated by calculation or extrapolated from neighbouring isotopes. A general feature of spectroscopic factors is that they are always quenched with respect to simple models by at least 50% (see figure 1.34) indicating the limitations of an independent-particle motion view of the nucleus [55].

1.2.11 Other approaches

The present discussion is not intended to be exhaustive but it is worth mentioning some further techniques which utilize theory to extract cross sections from indirect reaction studies, as well as emerging areas of theoretical modelling.

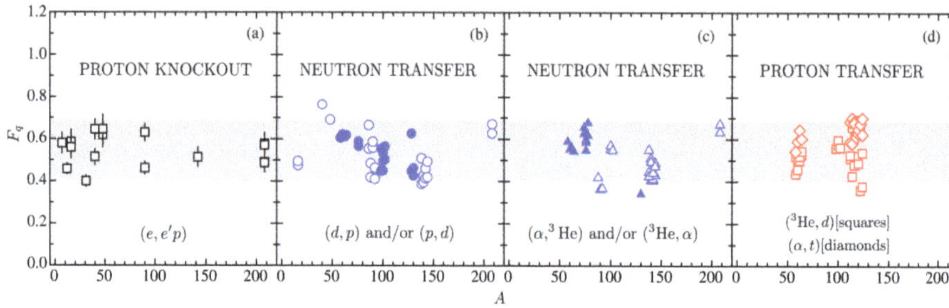

Figure 1.34. The quenching factor F_q versus target mass A for a wide variety of different classes of nuclear reaction. See [55] for a proper definition of the quenching factor; here, it can be taken as a proxy for the spectroscopic factor, B, discussed in the text. (Reproduced with permission from [55]. Copyright 2013 the American Physical Society.)

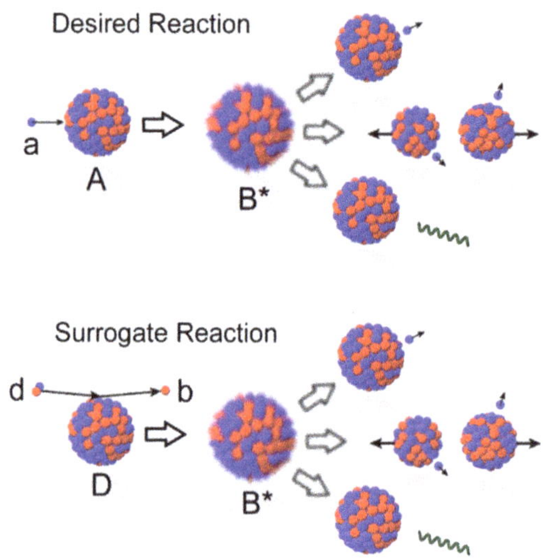

Figure 1.35. Schematic representation of the desired (top) and surrogate (bottom) reaction mechanisms. The basic idea of the surrogate approach is to replace the first step of the desired reaction, a + A, by an alternative (surrogate) reaction, d + D → b + B*, that is experimentally easier to access yet populates the same compound nucleus. The subsequent decay of the compound nucleus into the relevant channel, c + C, can then be measured and used to extract the desired cross section. Three typical decay channels are shown here: neutron evaporation, fission, and γ emission. (Reproduced with permission from [56]. Copyright 2012 the American Physical Society.)

1.2.11.1 The surrogate reaction method

The *surrogate reaction method* is a hybrid of theory and experiment which uses measurements of a reaction that proceeds through a particular compound nucleus stage to infer the cross section of a different, unmeasureable, reaction which proceeds through the same compound nucleus [56], as illustrated in figure 1.35.

The surrogate reaction used is typically a direct or inelastic scatter reaction, such as (d,p) or (p,p′), which leaves the compound nucleus of interest in an excited state. The decay of that compound nucleus is measured and used to fit model parameters which are then used in an optical model plus Hauser–Feshbach calculation of the desired reaction. It is assumed that the optical model calculation of the compound nucleus formation cross section will be relatively reliable, even for an unstable target nucleus, and so the surrogate method's focus is on constraining parameters for the Hauser–Feshbach portion of the calculation.

1.2.11.2 Trojan Horse method

The *Trojan Horse method* is somewhat similar to the surrogate method and has been used to study low energy charged particle reactions of relevance mostly to nuclear astrophysics but also fusion power [57]. The premise of the Trojan Horse method is to react two nuclei or a nucleus and a proton together, which would normally be repelled by one another due to the Coulomb interaction.

The Trojan Horse method replaces the desired, but unmeasurable reaction,

$$\mathrm{A} + x \rightarrow \mathrm{C} + \mathrm{c}, \tag{1.51}$$

with

$$\mathrm{A} + \mathrm{a} \rightarrow \mathrm{C} + \mathrm{c} + s, \tag{1.52}$$

where the replacement particle a is a composite particle made of $x + s$. A significant amount of theory is required to extract the cross section (or more often, the *astrophysical S-factor*) from the replacement reaction.

1.2.11.3 Ab initio *methods*

Ab initio calculations, where all potentials and interactions are modelled explicitly, can be performed for a variety of reactions and structure problems. However, the complexity of the calculation increases exponentially as the number of nucleons increases, meaning that the computational requirements can quickly become too great for the computing resources available. *Quantum algorithms*, i.e. calculations performed on a quantum computer, may provide, in the future, a route to accessing larger problem spaces than possible on current traditional computers. This is due to the problem space on a quantum computer scaling as 2^n (where n is the number of (quantum) bits, or qubits). A traditional computer can handle problems which simply scale as n.

1.2.11.4 Machine learning

Machine learning (ML) techniques show promise for predicting unmeasureable cross sections, or other nuclear data quantities, based on trends learned from other well understood nuclei. Many machine learning techniques also fall under the umbrella of *artificial intelligence* (AI).

1.3 Performing an evaluation

Evaluated nuclear data already exist for a wide range of isotopes, as well as for some isomeric states[13]. As discussed earlier, most evaluated files are part of large-scale libraries compiled by national and international working groups. These large libraries, such as ENDF/B (United States-led) [58], JEFF (European) [59], JENDL (Japanese) [60] and CENDL (Chinese) [61], contain data for hundreds of isotopes. Despite the wide range of nuclear data already available, new evaluations are often required. This may be due to older data requiring an update, possibly due to new calculations or experimental data being available, or a previous evaluation being too heavily focused on a particular reaction or energy region. For example, many evaluations were conducted to provide reliable data for traditional, i.e. fission, nuclear power applications, which do not require high quality data for higher energy reactions. This may mean that they are inadequate for future applications such as nuclear fusion power (see chapter 2) since the characteristic neutron energies from DT fusion are ~14 MeV whereas typical neutron spectra from fission reactors follow a broad energy distribution centred around ~1 MeV (see bottom panel of figure 1.22).

Most international, and national, working groups have high priority request lists where users can request evaluations for specific isotopes; this issue will be elaborated on later in section 5.2. Depending upon how niche the application is, however, a new evaluation may not be performed for some time. This section briefly describes the practical process of performing a nuclear data evaluation. It should be noted that there is no one-size-fits-all approach, but a typical methodology is described below. This can be seen as a key step on the nuclear data pipeline which ultimately translates laboratory measurement into a resource for real world applications.

1.3.1 Constraining theory against available data

The first step is to compile and assess all of the available experimental data for the isotope in question. An international open access database known as EXFOR contains the vast majority of experimentally measured cross section data. The existence of EXFOR simplifies the task of locating available data and the original documentation in which the measurements were reported.

The sifting of experimental data can be somewhat subjective. Often data are reported with only small, or even no, uncertainties. This cannot be taken at face value as there is no guarantee that uncertainties have been properly handled. The experimental set-up described in the original report should be assessed in order to determine whether the reported statistical and systematic uncertainties are justifiable. Data that appear suspect may be discarded or given a lower weight in any later comparison against theory.

[13] The definition of an isomer is subjective and often application-dependent (see section 1.1.2). For example, modelling a nanosecond long implosion of an inertial confinement fusion capsule will require far shorter-lived isotopes to be accounted for compared to a medical application where a radioactive isotope is observed in a patient for several hours.

Nuclear data evaluators from the US Cross Section Evaluation Working Group (CSEWG) project, which is responsible for delivering the ENDF/B library, have developed a methodology for applying templates to measured data with insufficient uncertainty information [62, 63]. These templates allow expected uncertainties, based on similar measurements, to be generated.

For calculations above the resonance region, a variety of codes exist which make use of a number of different reaction models. The open source code TALYS [31] is widely used by evaluators worldwide and combines the optical and Hauser–Feshbach models described in sections 1.2.7 and 1.2.8, along with a number of other methods to calculate nuclear data quantities. For resonance region calculations, it may be appropriate to employ *R*-matrix codes such as REFIT or SAMMY.

As previously discussed, for low energies (and low-mass isotopes), a sophisticated fitting procedure using an *R*-matrix code is performed when resonance data are available. For the case where no resonances are measured, a number of different approaches may be taken. These include: 'transplanting' resonances from a physically similar isotope; omitting resonances and having a completely smooth low energy cross section; creating a 'fake' resonance region with average properties matching those expected from an isotope with that mass, spin, etc; or creating a single artificial resonance which also replicates expected average parameters.

For higher energies, TALYS, or a similar code, can be used to generate cross sections in any reaction channel. Figure 1.36 shows the results of unadjusted TALYS calculations for neutrons incident on ^{89}Y compared to experimental data. The calculations can be repeated with adjusted input parameters until the match across all channels is within some tolerance.

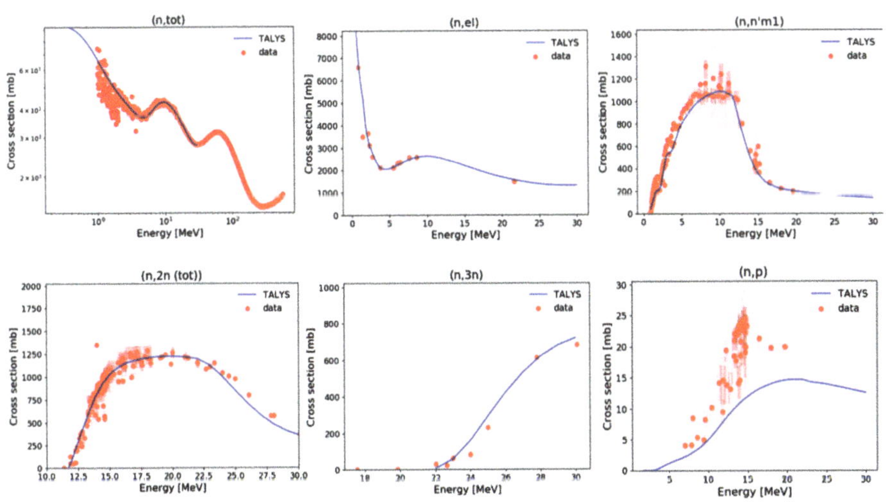

Figure 1.36. Example unadjusted calculations of n + ^{89}Y reaction cross sections, performed using the TALYS code, compared against experimental data taken from the EXFOR database.

1.3.2 Formats and processing

Once the nuclear data set generated is believed to be physically valid, it must be stored in an appropriate electronic format. The standard nuclear data format is ENDF(-6). ENDF is a legacy format based on an 80-character fixed format punch card approach. Figure 1.37 shows an extract of an example ENDF file.

Data in ENDF format can be onward processed into alternative formats which are compatible with different physics codes. The NJOY code is most commonly used for this onward processing, but others such as AMPX and PREPRO are also widely used. A more modern, more human-readable, xml-based format called GNDS is currently in development by an international collaboration and is expected to eventually replace the decades-old ENDF.

1.3.3 Grouped versus continuous data

In reality, a cross section continuously varies against incident neutron energy. However, storing and using a highly resolved set of cross section versus energy points may not be practical or possible in some applications. Cross sections are therefore often grouped into a set of discrete energy bins, with the cross section constant between the boundaries of each group. An example of the grouping/discretization of a cross section is illustrated in figure 1.38.

```
 1.400000+4 2.784400+1          0          0          0          01400 3 16
-8.473800+6-8.473800+6          0          0          1          121400 3 16
         12          2                                            1400 3 16
 8.778100+6 0.000000+0 1.000000+7 6.166000-3 1.100000+7 1.564000-21400 3 16
 1.200000+7 2.589000-2 1.300000+7 3.650000-2 1.400000+7 4.663000-21400 3 16
 1.500000+7 5.400000-2 1.600000+7 5.620000-2 1.700000+7 5.734000-21400 3 16
 1.800000+7 5.830000-2 1.900000+7 5.870000-2 2.000000+7 5.892000-21400 3 16
 0.000000+0 0.000000+0          0          0          0          01400 3  0
```

Figure 1.37. Example extract from an ENDF format nuclear data file.

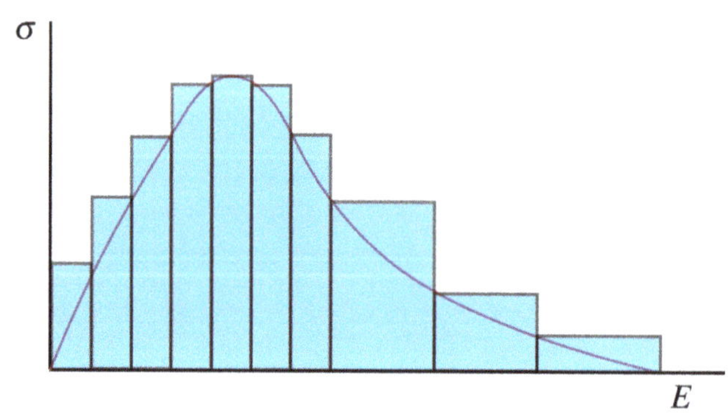

Figure 1.38. Comparison of a grouped cross section with the continuous underlying curve. Here a weighting spectrum has (incorrectly) not been used and the group cross sections are simply equal to the maximum value of the continuous cross section within each bin.

In order to determine the value of the cross section in each bin, a *weighting function* is typically used. The weighting function is a representative high-resolution neutron spectrum calculated for the nuclear system to which the grouped data will be applied[14]. This weighting function highlights the energy regions of importance within the problem and helps scale the cross section in each group to be equal to the continuous values at those points, rather than simply being an average of the continuous cross section between the boundaries of the group.

The use of weighting functions becomes more important as the energy group structure used becomes coarser. To condense a finely grouped or continuous cross section to a coarser group structure, a weighting spectrum should be used.

Between the range of energies bounding a coarse group, each fine group cross section σ_i must have a corresponding weight w_i. To apply this to the cross section in a single coarse energy group, which contains multiple finer groups, the weight/flux w for each fine group i within the coarser group is multiplied by the cross section of each fine group and summed to form a weighted cross section σ_w for the entire coarse group, i.e.

$$\sigma_w = \sum_i w_i \times \sigma_i. \tag{1.53}$$

The values of the weighting spectrum are arbitrarily scaled and are normalized when calculating the coarse group cross section. The final weighted cross section for the coarse group may then be calculated using

$$\sigma_{\text{coarse}} = \frac{\sigma_w}{\sum_i w_i}. \tag{1.54}$$

This process is repeated for each coarse group.

Figure 1.39 illustrates the benefits of using a weighting function to scale the grouped cross sections.

Using a grouped cross section structure can greatly simplify storing nuclear data and the use of it in modelling codes. GENDF (Grouped-ENDF) is one kind of grouped nuclear data format and (a slightly modified version) is utilized by the widely used inventory code FISPACT-II.

Some applications however require higher resolution data or are designed to use the 'true' physical cross section. Continuous, or *point-wise* nuclear data can also be provided for these cases. PENDF (Point-wise ENDF) and ACE (A Compact ENDF) are both examples of point-wise formats, with ACE utilized by the widely used MCNP transport code.

[14] A weighting function may also be derived from an existing parameterized equation.

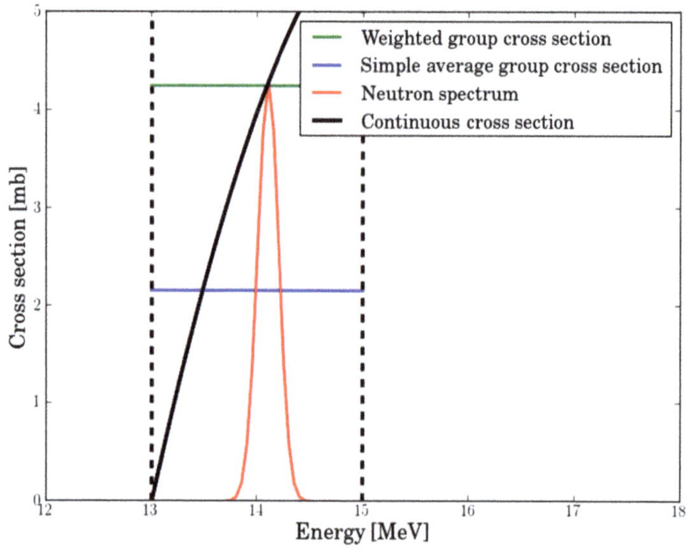

Figure 1.39. Illustration of the benefit of using a weighting function to scale grouped cross sections. For a single energy group (13–15 MeV) two different grouped cross sections are shown; the simple average (blue) which takes the mean value of the underlying cross section (black) between the two group boundaries and a weighted cross section (green) which gives a value more appropriate for the energy regions of interest in the model, as indicated by the neutron spectrum (red).

1.3.4 Angular distributions

For grouped nuclear data, the angular distributions of outgoing secondary particles are also stored in a reduced format. The continuous distribution, between 0° and 180°[15] is stored as a set of coefficients which may be used to reconstruct the distribution as a sum of Legendre polynomials.

For an initial angular distribution $f(y)$, where $y = \cos\theta$, the coefficient for each Legendre polynomial P_l can be calculated via

$$a_l = \frac{2l+1}{2}\int_{-1}^{+1} f(y)P_l(y)\mathrm{d}y, \tag{1.55}$$

where l here is the order of the polynomial rather than the angular momentum of a partial wave. The Legendre coefficients a_l allow the entire continuous angular distribution to be stored in a much reduced format within a nuclear data library. An angular distribution $\frac{\mathrm{d}\sigma}{\mathrm{d}\Omega}(y)$ may then be reconstructed in a modelling code using

$$\frac{\mathrm{d}\sigma}{\mathrm{d}\Omega}(y) \approx \sum_l \frac{2l+1}{2} a_l P_l(y). \tag{1.56}$$

[15] It is assumed that the distributions are azimuthally symmetric so storing a distribution between 180° and 360° is unnecessary.

The closeness of the match between $\frac{d\sigma}{d\Omega}(y)$ and $f(y)$ will depend upon the number of Legendre polynomials used in the reconstruction.

1.3.5 Benchmarking

After the evaluated data set has been processed into a format compatible with a physics code, benchmark calculations can be performed in order to test how it performs The type of benchmark used will depend upon the isotope under investigation. For example, for a fissile isotope such as uranium or plutonium, a *critical assembly* model can be used [64]. Figure 1.40 shows the Godiva-IV critical assembly which consists of a cylinder core comprising 65 kg of highly enriched uranium. The *criticality*, in terms of *k-effective*, is measured in a range of experiments which may be reproduced using a variety of modelling codes. *k-effective* gives a measure of how critical a mass of fissile material is; a value below 1 means the system is sub-critical and a value greater than 1 that the system is super-critical. This issue will be returned to in more detail in section 3.1.

The International Criticality Safety Benchmark Evaluation Project (ICSBEP) handbook [65] is a compilation of well modelled criticality benchmarks suitable for testing new nuclear data. Similarly, Shielding Integral Benchmark Archive and Database (SINBAD) [66] is a compilation of *shielding* benchmarks which model the emission of radiation from various nuclear facilities and experiments.

Different types of codes are appropriate for different kinds of benchmarks. For a benchmark model used to determine the distribution of neutrons in a facility or reactor, a neutron transport code would be appropriate. These codes may be one-, two- or three-dimensional depending upon the complexity of the model and the symmetries present. An example 3D transport code is MCNP which uses ACE-formatted nuclear data. For a benchmark to calculate the transmutation of an isotope due to a neutron flux, an *activation* code would be appropriate. An example of such a code is FISPACT-II which uses a (modified form of) GENDF-formatted data.

The main purpose of benchmark calculations is to test that the data can be retrieved and used by the required physics code and that it does not introduce any unexpected, or unphysical, results.

1.3.6 Uncertainties

Unfortunately, like any other microscopic quantity, cross sections (or other nuclear data quantities) cannot be measured exactly and it is unlikely that calculations for complex quantum systems will be perfect. Therefore, nuclear data need to be supplied with quoted uncertainties. Such nuclear data uncertainties are usually stored as covariance matrices. Covariance describes the correlated uncertainty between one energy point (or group/bin) and another. Example covariance matrices are given in figure 1.41.

The covariance of the cross section at one energy point with itself will simply equal the variance of the cross section at that point, i.e. the diagonal terms of the matrix are equal to the variance of the cross section as a function of energy. By performing many (hundreds or even thousands of) calculations for each reaction

Figure 1.40. A photograph of the Godiva-IV critical assembly; for details see [64]. (Reproduced from [64]. CC-BY-4.0.)

with input parameters varied between physical bounds, one can calculate covariances and convert those into a matrix format. The covariance of two discrete quantities x and y, e.g. a cross section in two different energy groups, can be given by

$$\text{cov}(x, y) = \frac{1}{n}\sum_i (x_i - \bar{x})(y_i - \bar{y}), \tag{1.57}$$

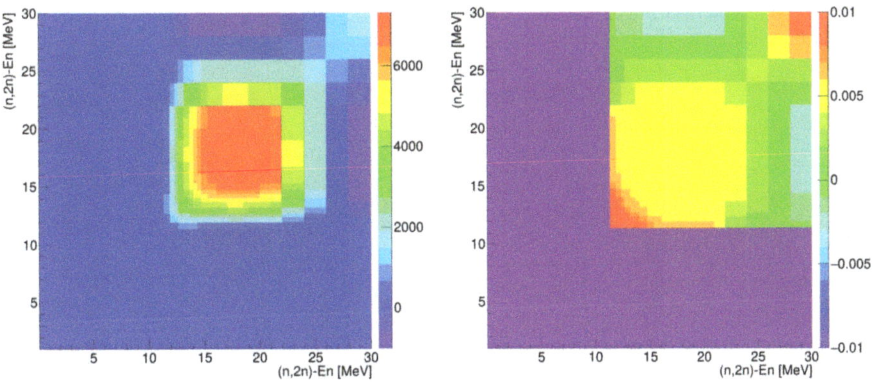

Figure 1.41. Example absolute and relative covariance matrices for the same (n,2n) reaction.

where n is the number of measurements or calculations of each quantity, x_i and y_i are the individual measured or calculated values, and $\bar{x}$ and $\bar{y}$ are the mean or expectation values of x and y, respectively.

Covariances created using equation (1.57) are *absolute* covariances. *Relative* covariances, where the covariances are expressed as a proportion of $\bar{x}$ and $\bar{y}$ are sometimes more useful and/or easier to interpret. Relative covariances may be created via

$$\mathrm{rel}(x, y) = \frac{\mathrm{cov}(x, y)}{\sqrt{\bar{x} \cdot \bar{y}}}. \tag{1.58}$$

Relative covariances are often stored in nuclear data files as, if desired, they may even be used in conjunction with cross sections taken from other libraries to provide rough uncertainty estimates.

1.4 Data beyond cross sections

Our focus so far has been on cross sections but there are a number of other aspects of nuclear data that should be considered. Here, we briefly describe a number of these with a particular emphasis on parameters which are discussed later in the book.

1.4.1 Average number of neutrons per fission event—$\bar{\nu}$

When a nucleus undergoes fission, in addition to the two (or possibly more) fragments (i.e. smaller nuclei) created, a number of other particles will also be emitted. The average number of neutrons emitted per fission event, $\bar{\nu}$, is a very important quantity for many nuclear applications. In a nuclear (fission) reactor, neutrons emitted from fission reactions are used to induce subsequent fission events and hence sustain a chain reaction and generate power at a steady level. Control rods containing neutron-absorbing isotopes are used to control the number of neutrons in the system. If the number of neutrons is too high, there will be a risk of a

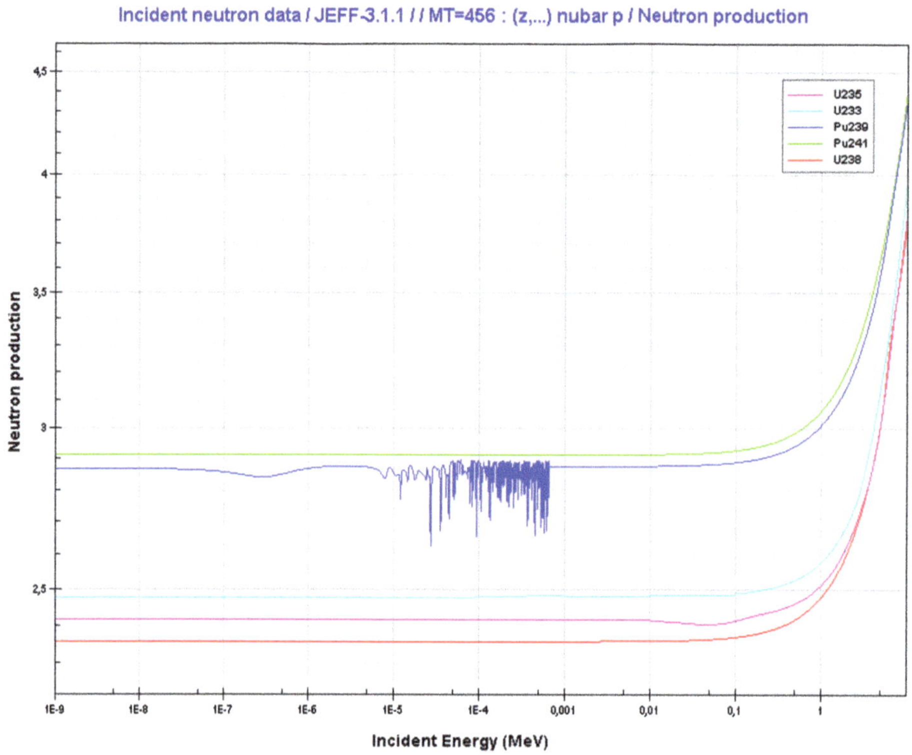

Figure 1.42. $\bar{\nu}$ values as a function of energy for a number of different actinide isotopes including $^{233,235,238}U$ and $^{239,241}Pu$. The plot was created using the JANIS tool from the NEA.

runaway chain reaction occurring, and if there are too few present there will be a risk of the chain reaction dying out; this is discussed from a more practical perspective in section 4.1.4. A knowledge of $\bar{\nu}$ is important in order to understand the response of the system to the presence of control rods and to therefore be able to sustain the chain reaction at its optimum level. $\bar{\nu}$ is an energy-dependent function that also varies by isotope. Figure 1.42 shows $\bar{\nu}$ values as a function of energy for a number of different isotopes.

1.4.2 Prompt fission neutron spectrum

In addition to $\bar{\nu}$, the energy spectrum of the neutrons produced by fission (the prompt fission neutron spectrum (PFNS)) is also incredibly important for understanding and predicting the performance of a nuclear system. Like $\bar{\nu}$, the shape of the PFNS also depends on the energy of the incident neutron which induced the fission reaction, as well as the type of target isotope. Determining the form of the PFNS for the major actinides at a sufficient range of incident neutron energies is a significant area of ongoing research due to the detailed physics involved and measurements required.

1.4.3 Fission product distributions

For fissile isotopes, information on the lighter fragments produced via their fissioning may be contained within a nuclear data library. These data are stored as distributions of the decay products as a function of Z and A. *Independent* fission yields give the distribution of products immediately following the fission of the parent actinide, i.e. before the products undergo further decays. *Cumulative* fission yields give the distribution following the decay of the products to their final stable states.

1.4.4 Energy release

Nuclear reactions may be endo- or exothermic. The *Q-value* for a reaction gives a measure of the energy absorbed or liberated. The energy available to the products of a reaction will depend upon the incident energies and the specific Q-value. As these reaction products travel through a material, energy will be deposited due to interactions with the material present through different physical process. The energy deposited will depend strongly on the particle type. For example, neutrons lose energy discontinuously through nuclear reactions and scattering, while charged particles deposit energy continuously along their track (see appendix A for discussion of the physics of energy loss by charged particles in matter). When, for example, nuclei within the material gain kinetic energy through interactions with neutrons, the material will be heated. The heating of materials due to nuclear reactions is referred to as either nuclear heating or more commonly as *energy release*. If all particle types are explicitly tracked through a calculation, then each interaction will be modelled and the energy deposited throughout each material can be determined via a knowledge of reaction Q-values and transferred kinetic energies. However, most modelling codes do not include the tracking of all particle types. 'Kinetic energy release in materials' (kerma) coefficients are used to help account for heating due to particles which are not fully included in calculations.

Kerma coefficients $k_{ij}(E)$ are (incident particle) energy dependent and differ for each target material i and reaction type j. The coefficients act in a similar way to cross sections, but with units of energy × cross section (eV · barns). Unlike reaction Q-values, kerma coefficients are not stored in nuclear data libraries as they must be created for specific use cases, i.e. they depend upon which particle types are modelled in the code being used. If, for example, a modelling code simulates the transport and interactions of neutrons, but does not include photons or charged particles, then kerma coefficients must be created for reactions which generate photons or charged particles, e.g. (n,γ) or (n,p). When one of these reactions occurs in the calculation, the kerma coefficient will be used to locally heat around the target nucleus. Obviously this local heating is an approximation as, in reality, γ-rays emitted may deposit their energy over a greater distance, however, it ensures that the effect of these secondary particles on the material is not neglected.

1.4.5 Thermal scattering law data

At very low (below a few eV) energies, neutron elastic scattering cross sections become sensitive to the chemical structure and material the target nucleus is part of, e.g. cross sections of carbon will be different if the carbon is bound in diamond compared to graphite. Molecular excitations and phonons within the material will imprint themselves on the cross section and so the phase of the target material and its chemical composition needs to be known. For example, at these energies, water cannot be treated as independent oxygen and hydrogen nuclei and so must have its own scattering data, referred to as *thermal scattering law* (TSL), or sometimes $S(\alpha, \beta)$, data.

In, for example, a nuclear reactor, neutrons are intentionally moderated, or elastically scattered, to very low energies where they are unable to escape the reactor vessel and also have a high fission cross section. Good quality TSL data for a range of moderating materials are therefore required in order to accurately model the operation of a reactor core. TSL data are stored as a range of $S(\alpha, \beta)$ values where α and β are dimensionless momentum and energy transfer variables. These $S(\alpha, \beta)$ values are used in standard formulae to determine the double differential cross section—the cross section as a function of both change in angle and energy.

References

[1] Jenkins D G and Wood J L 2021 *Nuclear Data: A Primer* (IOP Publishing)

[2] Jenkins D G and Wood J L 2023 *Nuclear Data: A Collective Motion View* (IOP Publishing)

[3] Jenkins D G and Wood J L 2024 *Nuclear Data: An Independent-Particle Motion View* (IOP Publishing)

[4] Mayer M G 1948 On closed shells in nuclei *Phys. Rev.* **74** 235

[5] Goeppert-Mayer M 1969 On closed shells in nuclei. II *Phys. Rev.* **75** 1969

[6] Haxel O, Jensens J H D and Suess H E 1949 On the "magic numbers" in nuclear structure *Phys. Rev.* **75** 1766

[7] Bentley M A 2022 Excited states in isobaric multiplets—experimental advances and the shell model approach *Physics* **4** 995–1011

[8] Zelevinsky V and Horoi M 2018 Nuclear level density, thermalization, chaos and collectivity *Prog. Part. Nucl. Phys.* **105** 180–213

[9] Chadwick J 1914 The intensity distribution in the magnetic spectrum of β particles from radium (B+C) *Verh. Phys. Gesell.* **16** 383–91

[10] Reines F and Cowan C L 1953 Detection of the free neutrino *Phys. Rev.* **92** 830

[11] Reines F and Cowan C L 1959 Free neutrino absorption cross section. I. Measurement of the free neutrino absorption cross section by protons *Phys. Rev.* **113** 273

[12] Fermi E 1934 Versuch einer Theorie der Beta-Strahlen *Z. Phys.* **88** 161–71

[13] Rubio B and Gelletly W 2009 *Euroschool Lectures on Physics with Exotic Beams* vol 3 (Lecture Notes in Physics vol 764) (Springer)

[14] Rubio B, Gelletly W and Naviliat-Cuncic O 2022 Beta decay: probe for nuclear structure and the weak interaction *Handbook of Nuclear Physics* (Springer)

[15] Turkat S, Mougeot X, Singh B and Zuber K 2023 Systematics of log*ft* values for β^- and EC/β· transitions *At. Data Nucl. Data Tables* **152** 101584

[16] Christenson J H, Cronin J W, Fitch V L and Turlay R 1964 Evidence for the 2π decay of the K_2^0 meson *Phys. Rev. Lett.* **13** 138
[17] Wu C S, Ambler E, Hayward R W, Hoppes D D and Hudson R P 1957 Experimental test of parity conservation in beta decay *Phys. Rev.* **105** 1413
[18] Batchelder J C 2020 Recommended values for β-delayed proton and α emission *At. Data Nucl. Data Tables* **132** 101323
[19] Bender M *et al* 2020 Future of nuclear fission theory *J. Phys. G: Nucl. Part. Phys.* **47** 113002
[20] Sublet J-Ch, Eastwood J W, Morgan J G, Gilbert M R, Fleming M and Arter W 2017 FISPACT-II: An advanced simulation system for activation, transmutation and material modelling *Nucl. Data Sheets* **139** 77–137
[21] Cockcroft J D and Walton E T S 1932 Experiments with high velocity ions II: The disintegration of elements by high velocity protons *Proc. R. Soc.* A **137** 229–42
[22] Rolfs C and Rodney W S 1988 *Cauldrons in the Cosmos* (University of Chicago Press)
[23] Iliadis C 2007 *Nuclear Physics of Stars* (Wiley)
[24] Thompson I J and Nunes F M 2012 *Nuclear Reactions for Astrophysics* (Cambridge University Press)
[25] Zilges A, Balabanski D L, Isaak J and Pietralla N 2022 Photonuclear reactions—from basic research to applications *Prog. Part. Nucl. Phys.* **122** 103903
[26] Shusterman J A *et al* 2019 The surprisingly large neutron capture cross-section of ^{88}Zr *Nature* **565** 328
[27] Block R C, Danon Y, Gunsing F and Haight R C 2010 *Handbook of Nuclear Engineering* (Springer)
[28] Hodgson P E 1987 Compound nucleus reactions *Rep. Prog. Phys.* **50** 1171–228
[29] Colonna N, Gunsing F and Käppeler F 2018 Neutron physics with accelerators *Prog. Part. Nucl. Phys.* **101** 177–203
[30] Ganesapandy T S, Jeremiah J J, Dahiwale S S, Dhike S D and Bhoraskar V N 2019 Analysis of neutron induced (n,γ) and (n,2n) reactions on ^{232}Th from reaction threshold to 20 MeV *Appl. Radiat. Isot.* **150** 70–8
[31] Koning A, Hilaire S and Goriely S 2023 TALYS: modeling of nuclear reactions *Eur. Phys. J.* A **59** 131
[32] Heyde K and Wood J L 2019 *Quantum Mechanics for Nuclear Structure: A Primer* (IOP Publishing)
[33] Heyde K and Wood J L 2020 *Quantum Mechanics for Nuclear Structure: An Intermediate Level View* (IOP Publishing)
[34] Singh J 2016 New approaches to the physics of weakly-bound nuclei: treatmentz of continuum in ^{6}He *PhD Thesis* University of Padova
[35] Piyadasa R A D, Kawai M, Kamimura M and Yahiro M 1999 Convergence of the solution of the continuum discretized coupled channels method *Phys. Rev.* C **60** 044611
[36] Moro A M, Arias J M, Gómez-Camacho J, Martel I, Pérez-Bernal F, Crespo R and Nunes F 2002 Coupling to breakup channels using a transformed harmonic oscillator basis *Phys Rev.* C **65** 011602
[37] Austern N, Iseri Y, Kamimura M, Kawai M, Rawitscher G H and Yahiro M 1987 Continuum-discretized coupled-channels calculations for three-body models of deuteron-nucleus reactions *Phys. Rep.* **154** 125
[38] Yahiro M, Nakano N, Iseri Y and Kamimura M 1982 Coupled-discretized-continuum-channels method for deuteron breakup reactions based on three-body model: Justification of

the method for truncation and discretization of the p–n continuum *Prog. Theor. Phys.* **67** 1467
[39] Kamimura M, Yahiro M, Iseri Y, Sakuragi Y, Kaeyama H and Kawai M 1986 Chapter I. Projectile breakup processes in nuclear reactions *Prog. Theor. Phys. Suppl. Ser.* **89** 1
[40] Yahiro M and Kamimura M 1981 Coupled channel method of deuteron breakup reactions based on totally antisymmetrized three-body model: a new method for calculating breakup form factors *Prog. Theor. Phys.* **65** 2046
[41] Rasoanaivo R Y and Rawitscher G H 1989 Discretization methods of the breakup continuum in deuteron–nucleus collisions *Phys Rev.* C **39** 1709
[42] Matsumoto T, Kamizato T, Ogata K, Iseri Y, Hiyama E, Kamimura M and Yahiro M 2003 New treatment of breakup continuum in the method of continuum discretized coupled channels *Phys Rev.* C **68** 064607
[43] Pérez-Bernal F, Martel I, Arias J M and Gómez-Camacho J 2002 A new basis set for continuum discretization *Few-Body Syst.* **13** 217–24
[44] Moro A M, Arias J M, Gómez-Camacho J and Pérez-Bernal F 2009 Analytical transformed harmonic oscillator basis for continuum discretized coupled channels calculations *Phys. Rev.* C **80** 054605
[45] Matsumoto T, Hiyama E, Yahiro M, Ogata K, Iseri Y and Kamimura M 2004 Four-body CDCC analysis of ^{6}He+^{12}C scattering *Nucl. Phys.* A **738** 471
[46] Rodríguez-Gallardo M, Arias J M, Morgan J G, Johnson R C, Moro A M, Thompson I J and Tostevin J A 2008 Four-body continuum-discretized coupled-channels calculations using a transformed harmonic oscillator basis *Phys. Rev.* C **77** 064609
[47] Singh B P, Sharma M K and Prasad R 2022 *Pre-equilibrium Emission in Nuclear Reactions* (IOP Publishing)
[48] Hodgson P E 1971 The nuclear optical model *Rep. Prog. Phys.* **34** 765
[49] Dickhoff W H and Charity R J 2019 Recent developments for the optical model of nuclei *Prog. Part. Nucl. Phys.* **105** 252–99
[50] Hauser W and Feschbach H 1952 The inelastic scattering of neutrons *Phys. Rev.* **87** 366
[51] Hammache F and de Séréville N 2021 Transfer reactions as a tool in nuclear astrophysics *Front. Phys.* **8** 602920
[52] Krane K S 1988 *Introductory Nuclear Physics* (Wiley)
[53] Lane A M and Thomas R G 1958 R-matrix theory of nuclear reactions *Rev. Mod. Phys.* **30** 257
[54] Vogt E 1962 Theory of low energy nuclear reactions *Rev. Mod. Phys.* **34** 723
[55] Kay B P, Schiffer J P and Freeman S J 2013 Quenching of cross sections in nucleon transfer reactions *Phys. Rev. Lett.* **111** 042502
[56] Escher J E, Harke J T, Dietrich F S, Scielzo N D, Thompson I J and Younes W 2012 Compound-nuclear reaction cross sections from surrogate measurements *Rev. Mod. Phys.* **84** 353
[57] Tumino A *et al* 2025 Indirect methods with transfer reactions: the Trojan Horse method and the asymptotic normalization coefficient *Prog. Part. Nucl. Phys.* **143** 104164
[58] Chadwick M B *et al* 2011 ENDF/B-VII.1 nuclear data for science and technology: cross sections, covariances, fission product yields and decay data *Nucl. Data Sheets* **112** 2887–996
[59] Plompen A J M *et al* 2020 The joint evaluated fission and fusion nuclear data library, JEFF-3.3 *Eur. J. Phys.* A **56** 181

[60] Nakagawa T *et al* 1995 Japanese evaluated nuclear data library version 3 Revision-2: JENDL-3.2 *J. Nucl. Sci. Technol.* **32** 1259–71

[61] Ge Z *et al* 2020 CENDL-3.2: The new version of Chinese general purpose evaluation nuclear data library *EPJ Web Conf.* **239** 09001

[62] Neudecker D *et al* 2023 Templates of expected measurement uncertainties *EPJ Nucl. Sci. Techol.* **9** 35

[63] Lewis A *et al* 2023 Template of expected measurement uncertainties for total neuton cross-section observables *EPJ Nucl. Sci. Techol.* **9** 34

[64] Thompson N W *et al* 2023 The National Criticality Experiments Research Center and its role in support of advance reactor design *Front. Energy Res.* **10** 1082389

[65] Bess J D, Marshall M A, Gulliford J and Hill I 2016 International handbook of evaluated criticality safety benchmark experiments *Technical report* Nuclear Energy Agency/OECD

[66] Kodeli I A and Sartori E 2021 SINBAD—Radiation shielding benchmark experiments *Ann. Nucl. Energy* **159** 108254

[67] Jenkins D 2020 *Radiation Detection for Nuclear Physics* (IOP Publishing)

[68] Rowe D J and Wood J L 2010 *Fundamentals of Nuclear Models: Foundational Models* (World Scientific)

[69] Jain A K, Maheshwari B, Garg S, Patial M and Singh B 2015 Atlas of nuclear isomers *Nucl. Data Sheets* **128** 1–130

Chapter 2

Nuclear data for fusion power

Greg Bailey, Mark Gilbert, Lee Packer and Alex Valentine

The race is on to develop nuclear fusion as a future source of energy. Realizing fusion on an industrial scale, however, will mean surmounting scientific and engineering challenges in a broad range of disciplines from plasma physics to material science. Nuclear data are also important since, for example, they play a crucial role in understanding the extent to which materials in a fusion reactor will be activated. A second aspect strongly related to nuclear data is tritium breeding—a technology that will continuously produce the fuel needed for a fusion reactor. Here, we provide an overview of the engineering design of a fusion reactor before highlighting some of the key areas where nuclear data are most relevant and where new data are needed.

Concepts: Fusion, neutron transport, activation, tritium breeding, neutron shielding, material damage

2.1 Introduction to fusion reactors

Energy is released when fusing together lighter nuclei to form a heavier, but more tightly bound, nucleus (see figure 1.2). Nuclear fusion occurs naturally in the cores of stars where the energy released counteracts gravitational collapse. Fusion reactors plan to harness the energy released from nuclear fusion to produce sustainable, low-carbon energy on Earth. The reactions occurring inside stars (which can fuse nuclei from H up to Fe and Ni) are slow, with the colossal amount of energy released being a product of a star's enormous size. This is the reason that terrestrial fusion reactors will need to exploit different reactions to achieve viable fusion power plants.

2.1.1 Fusion reactions and the choice of fusion fuel

For fusion to occur, the reacting nuclei must be physically close to one another, i.e. overcome the Coulomb barrier. All isotopes of hydrogen have a single charged component, meaning that the Coulomb barrier for hydrogen fusion is the lowest possible, implying that the fusion of hydrogen isotopes should require the least

doi:10.1088/978-0-7503-5102-7ch2 © IOP Publishing Ltd 2025. All rights, including for text and data mining (TDM), artificial intelligence (AI) training, and similar technologies, are reserved.

energy input for fusion to occur. Hydrogen is also the most abundant element in the Universe, so it is reasonable to explore hydrogen fusion reactions as an energy source.

The simplest form of hydrogen fusion, and fusion more generally, is the combination of two protons (^{1}H nuclei) to form a ^{2}H nucleus, otherwise known as the deuteron (d),

$$\mathrm{p} + \mathrm{p} \rightarrow \mathrm{d} + \mathrm{e}^+ + \nu_e, \quad Q = 1.442 \text{ MeV} \tag{2.1}$$

where Q is the reaction Q-value—the energy released/absorbed in the reaction (Q may be negative). In the example shown, $Q = 1.442$ MeV, which will be shared as kinetic energy between the reaction products. At stellar temperatures, the p–p fusion reaction requires *quantum tunnelling* to overcome the Coulomb barrier which results in lower reaction rates. For fusion power to be viable, higher reaction rates are needed, so the fuel will be hotter than typical H-burning stellar cores so that the reactants have high enough kinetic energy to overcome the Coulomb barrier. Unfortunately, this will require a relatively large amount of initial input energy. If large amounts of energy need to be input to allow for continuous fusion, the reaction exploited needs to provide the maximum energy output. For power generation, more energy needs to be output than input and the majority of the liberated energy should ideally be carried by a particle which can readily interact with matter so that the energy generated may be captured. There are a range of possible fusion reactions involving different permutations of the isotopes of hydrogen (see equation (2.2a)). The energy released in these reactions varies quite significantly. Indeed, not all such reactions result in energy release; for example, the fusion of a proton and tritium (equation (2.2d)) requires the absorption of 0.8 MeV in energy, so is not viable for energy production.

$$\mathrm{p} + \mathrm{d} \rightarrow {}^3\mathrm{He} + \gamma, \quad Q = 5.49 \text{ MeV} \tag{2.2a}$$

$$\mathrm{d} + \mathrm{d} \rightarrow \mathrm{t} + \mathrm{p}, \quad Q = 4.03 \text{ MeV}, 50\% \tag{2.2b}$$

$$\rightarrow {}^3\mathrm{He} + \mathrm{n}, \quad Q = 3.27 \text{ MeV}, 50\% \tag{2.2c}$$

$$\mathrm{p} + \mathrm{t} \rightarrow {}^3\mathrm{He} + \mathrm{n}, \quad Q = -0.8 \text{ MeV} \tag{2.2d}$$

$$\mathrm{d} + \mathrm{t} \rightarrow \alpha + \mathrm{n}, \quad Q = 17.6 \text{ MeV} \tag{2.2e}$$

$$\mathrm{t} + \mathrm{t} \rightarrow \alpha + 2\mathrm{n}, \quad Q = 11.3 \text{ MeV}. \tag{2.2f}$$

The energy, and therefore temperature, dependence of the cross sections for these reactions needs to be considered so that the most suitable fuel can be decided upon. Figure 2.1 shows the cross section for a selection of hydrogen fusion reactions as a function of centre of mass energy; the data are taken from the TENDL2019 nuclear

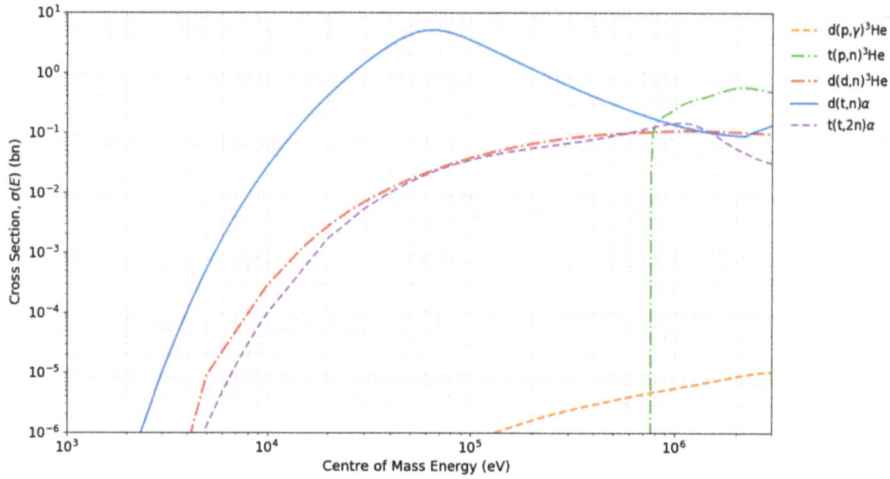

Figure 2.1. Cross sections for various fusion reactions involving different isotopes of hydrogen. The fusion of deuterium and tritium has the highest cross section and peaks at the lowest energy. These features make it the ideal choice for fusion reactors. All data are taken from the TENDL2019 nuclear data cross section library [1]. The legend uses nuclear reaction nomenclature, i.e. the notation $A(a, b)B$ means that particle a is incident on nucleus A producing nucleus B and emitting particle b (see section 1.2.1).

data cross section library [1]. The cross section for the fusion of deuterium and tritium[1] (solid blue line in figure 2.1) peaks at a cross section of around 3 b at a centre of mass energy of approximately 60 keV. Of all the reactions shown in equation (2.2a), this reaction would require the least energy input to obtain the maximum energy return. Indeed, DT-fusion has the highest Q-value of 17.6 MeV of the reactions considered. Here, the energy split between the products is as follows: 3.5 MeV for the heavier partner, the α particle, and 14.1 MeV for the lighter partner, the neutron. This makes DT-fusion an attractive choice to build a reactor concept around. Extracting energy from the high energy neutrons released is the basis of most current fusion reactor concepts.

Given that all of the reactions considered in figure 2.1 have the same Coulomb barrier (as both reactions have the same charge) it may be surprising that the DT-fusion cross section is so large and peaked at low energy. This can be explained by studying the structure of the intermediate and highly unstable ^{5}He nucleus created. A partial level scheme for ^{5}He is given in figure 2.2, which shows that a $\frac{3}{2}^+$ state is present at 16.84 MeV. Called the *Bretscher state* [4], named after Egon Bretscher who (alongside Anthony French) first discovered its existence [3], this state is very close in energy to the d + t separation energy of 16.92 MeV. As such, the five-nucleon system will readily occupy this resonance, leading to an increased fusion cross section. The Bretscher state almost always decays via neutron emission, but

[1] DT-fusion has been studied for a long time. It was first observed experimentally by Ruhlig in 1938 [2]; direct measurements were declassified and published in 1949 [3].

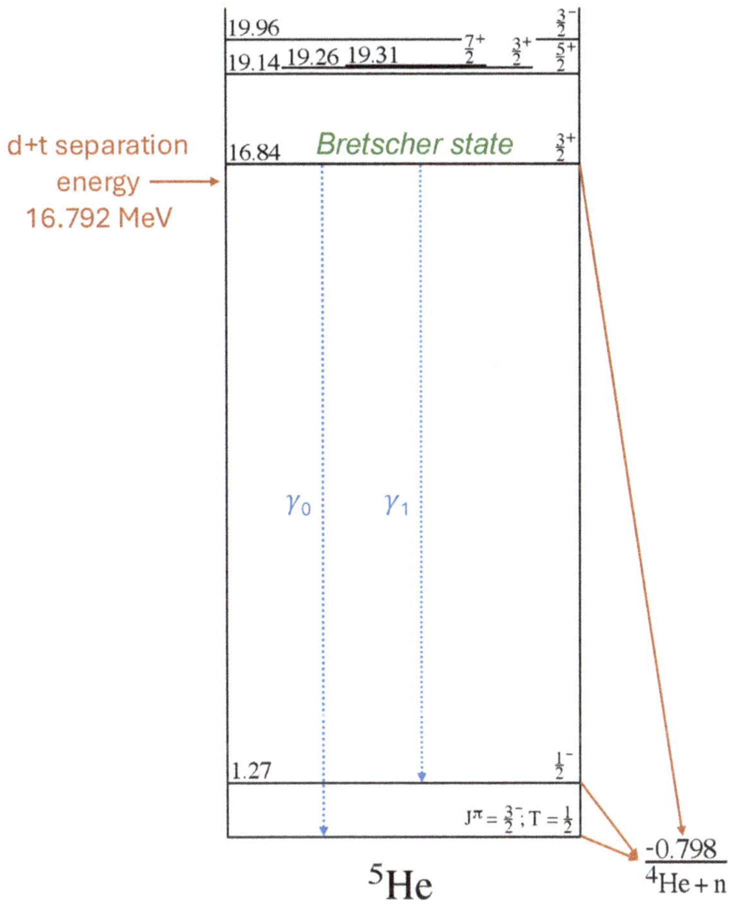

Figure 2.2. Low energy level scheme of ^{5}He highlighting the Bretscher state and the fusion-relevant decay pathways. The data presented in the level scheme are taken from [6, 7].

can very rarely undergo γ decay to one of the lower lying states which will themselves decay via neutron emission. The emitted γ-rays are very high in energy (~17 MeV) but are only emitted in approximately 1 in 10000 DT-reactions. Nevertheless, researchers are exploring whether detection of this high energy gamma-ray could be used to monitor the operation of a fusion reactor [5].

The major disadvantage of DT-fusion is the requirement to use tritium. Tritium is unstable (half-life of 12.33 years) and has a very low natural abundance being mostly produced from cosmic ray interactions with nitrogen in the upper atmosphere. Since a supply of tritium is required to fuel a DT-fusion device, solutions will need to be found to produce it artificially. One proposed solution is *tritium breeding*, using the neutrons released in DT-fusion to produce tritium via reactions on other nuclei. If this can occur readily enough it may be possible to achieve self-sufficiency in tritium; tritium breeding will be covered in greater detail in section 2.3.

2.1.2 Types of fusion reactors

As has been discussed, a fusion reactor exploiting DT-fusion for energy production will require the reaction to occur at a centre of mass energy of approximately 60 keV. To achieve these energies, the deuterium–tritium fuel mix will need to be heated to temperatures so great that the gas will become a plasma[2]. This plasma will be hotter than some stellar interiors and can easily damage reactor components/materials. The first hurdle for fusion reactors therefore is how to produce and contain such a plasma. Fusion reactor development has generally settled on one of two possible approaches: magnetic confinement and inertial confinement.

2.1.2.1 Magnetic confinement

Plasmas are mixtures of charged components so confining them using a magnetic field is an attractive solution. Powerful electromagnets can be used to shape and contain a plasma. The plasma can then be heated to a temperature where fusion reactions occur. If the confinement can be maintained, continuous operation of the device and therefore continuous energy production is possible. Due to the Lorentz force law[3], charged particles moving in a magnetic field will have curved orbits. Accordingly, the plasma chambers of magnetically confined fusion reactors are typically rings. An advantage of this configuration is that the charged particles can have effectively infinite path lengths within the plasma as they will have circular orbits. This allows the plasma to be contained as the particles should not impinge onto the containment vessel. In reality, fusion plasmas are turbulent and complex, for example, the motion of the charged particles themselves will induce a magnetic field, so particles will have helical orbits. Many magnets will be required to shape and contain such plasmas.

An advantage of magnetic confinement is that high energy neutrons will be able to escape the plasma as they are not charged. The neutrons will impinge on the structure surrounding the reactor, heating it. This heat can then be extracted from the coolant for electricity production. As they are charged, the α particles produced will be trapped in the plasma where they can impart their energy, aiding plasma heating.

The most common configuration of a magnetically confined fusion reactor is a *tokamak*. These are toroidal chambers surrounded by magnetic coils with a central solenoid which holds and shapes the plasma within the vessel while ensuring it does not touch the vessel. The word 'tokamak' derives from a Russian acronym meaning *toroidal chamber with magnetic coils*. Tokamaks were initially developed in the 1950s in Russia; recent examples include the Joint European Torus (JET) at Culham, UK, and ITER at Cadarache, France, which is hoping to be the first magnetic fusion device to produce net energy. Figure 2.3 shows the JET tokamak with the toroidal plasma chamber and shaping magnets highlighted.

[2] Plasma is a state of matter where the atoms are fully ionized. The electrons and ions are separated.

[3] $\mathbf{F} = q(\mathbf{E} + \mathbf{v} \times \mathbf{B})$ where $\mathbf{F}$ is the force experienced, q is the charge, $\mathbf{E}$ is the electric field, $\mathbf{B}$ the magnetic field and $\mathbf{v}$ is the velocity.

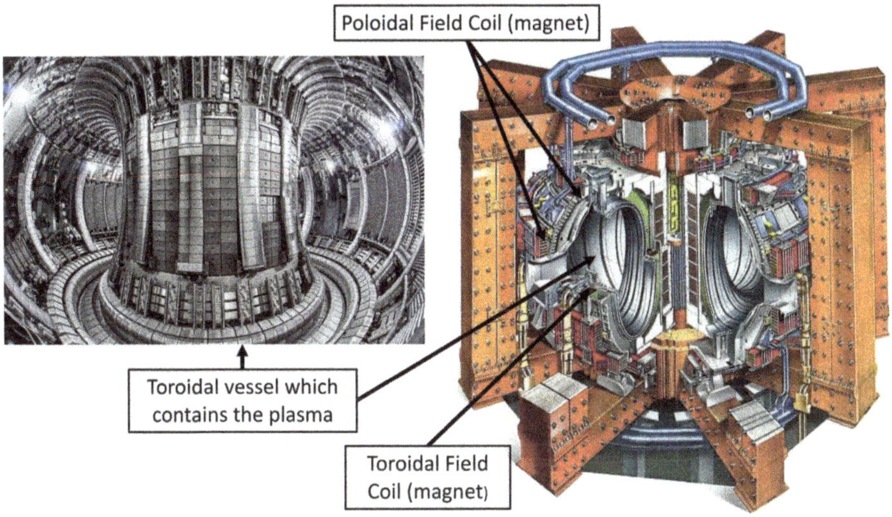

Figure 2.3. The right-hand image is a cartoon of the JET tokamak with the magnet required to hold and shape the plasma labelled. The left image shows the interior of the JET vessel. (This [JET vessel internal view] image and [JET cutaway drawing 1980] image have been obtained by the author(s) from the Wikimedia website where they were made available by [Geek3] under a CC BY 4.0 licence. They are included within this chapter on that basis. They are attributed to [EUROfusion].)

An alternative to the tokamak is the stellarator. These devices have twisted plasma chambers without a central solenoid. These devices embrace the helical nature of the plasma orbit, giving better confinement but requiring highly complex magnetic coil configurations. The engineering challenges associated with the construction of stellarators has made the tokamak the more popular choice.

2.1.2.2 Inertial confinement

Instead of attempting to continuously contain and heat a plasma, fusion could be initiated instantaneously by rapid compression, therefore rapid heating, of a fuel-rich target. If the outer layer of a spherical fuel pellet is rapidly heated the outer layer will explode outwards. Due to Newton's third law, the fuel contained within will collapse inward, the resulting shock-wave causing the compressed fuel to reach the densities and temperatures required for fusion. Once this is achieved, fusion occurs, neutron are emitted and the plasma bubble cools via expansion. This process is shown in figure 2.4.

Similar to the magnetic confinement scenario, the heavy α particles will impart their energy locally, due to their short mean free path in the dense compressed fuel and aid fuel heating. The energy of the emitted neutrons is planned to be captured in a similar manner to magnetically confined systems—the cooling of materials heated by neutron bombardment. The pellets are typically heated directly by

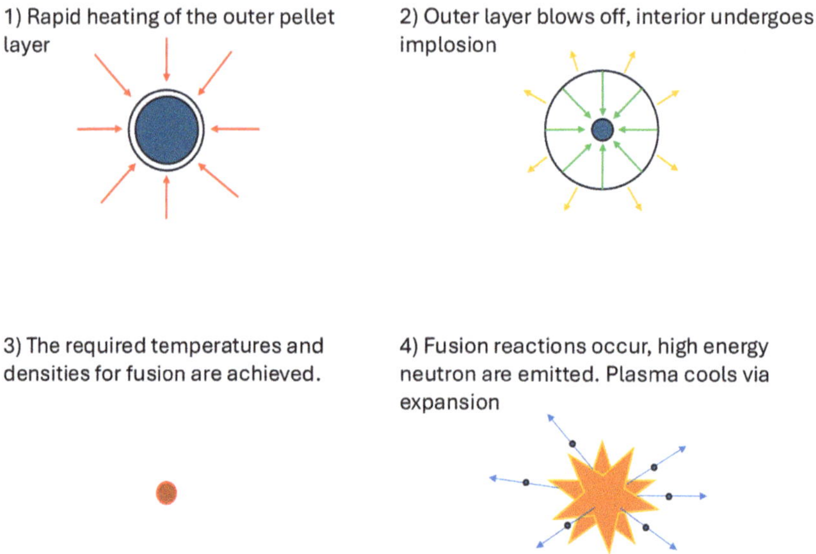

Figure 2.4. Cartoon of the process of inertial confinement fusion. Here, the external heating may be direct or indirect; the complete process occurring within a hohlraum.

powerful lasers or indirectly by x-rays produced from lasers irradiating the interior surface of a cylindrical hohlraum[4] made of heavy metal which encapsulate the fuel pellet.

At the time of writing, inertial confinement fusion (ICF) has produced the largest fusion energy gain. These experiments took place at the National Ignition Facility (NIF), Lawrence Livermore National Laboratory, US, which claimed to have produced 3.15 MJ after supplying 2.05 MJ of laser energy [8]. This is the first scientific break-even from a fusion device, although 300 MJ was required to produce the 2.05 MJ input. NIF consists of a spherical target chamber with a number of high power lasers focused on the target. These are directed into a gold-plated hohlraum which contains the fuel pellet. To exploit such a configuration for consistent energy production, several shots will have to occur per second, likely with many target chambers operating simultaneously.

2.1.2.3 Aneutronic fusion

While most fusion reactor designs are focusing on DT-fusion for the reasons outlined previously, other approaches are being explored. As will be discussed in later sections, fusion neutron sources have many challenging aspects and fuels which do not produce any neutrons are being explored as an alternative. The currently most favoured aneutronic fusion reactions are

$$\mathrm{d} + {}^{3}\mathrm{He} \rightarrow \alpha + \mathrm{p}, \qquad Q = 18.3\ \mathrm{MeV} \tag{2.3a}$$

[4] A hohlraum is a cavity where the bounding wall is in radiative equilibrium with the radiant energy within.

$$p + {}^{11}B \rightarrow 3\alpha, \qquad Q = 8.7 \text{ MeV}. \tag{2.3b}$$

^{3}He only makes up $2 \times 10^{-4}\%$ of natural He, so is not a practical fuel. Accordingly, most effort in terms of aneutronic fusion focuses on the p–^{11}B system. The greatest challenge in exploiting such a reaction is the extremely high energy required for them to occur due to the increased Coulomb barrier. These require at least an order of magnitude higher temperatures than DT-fusion. These reactions only produce charged particles so extracting the energy from a plasma requires different approaches to DT-fusion. While experiments continue, there are currently no operating reactors or devices using aneutronic fusion.

The remainder of this chapter will focus exclusively on DT-fusion devices. While aneutronic fusion and other alternative fuel may be viable in the future, the first generation of fusion reactors will use neutron-releasing DT-fusion.

2.2 Radiation fields in fusion reactors

2.2.1 Introduction to radiation transport

Understanding and predicting radiation fields emitted from a given source and present throughout a reactor geometry is an essential component of nuclear engineering and developing novel nuclear applications. Radiation or particle transport methods calculate such fields by modelling a particle's trajectory through a given geometry and materials. When designing a fusion reactor, knowledge of the neutron and photon spectra across the complete reactor geometry provides insight into tritium breeding (see section 2.3), shielding requirements (Section 2.4) and materials evolution (section 2.5). Fusion is most often concerned with neutron transport (or neutronics) and photon transport.

When a neutron enters a material, it may interact with the nuclei contained within. The multitude of interactions which a neutron can undergo will each impact the particle's trajectory and energy. Some events will lower the neutron energy (scattering), others will create more neutrons (multiplication), some will remove neutrons (absorption) and some will produce secondary particles. Figure 2.5 shows a cartoon of this process. To predict the spectrum at a given position, the probability of each interaction occurring is needed. This information is provided by nuclear data libraries; reaction cross sections provide the interactions and their energy-dependent probabilities, as well as the angular distributions of the outgoing particles. From this data, the radiation fields around a given source can be predicted from the statistical likelihood of certain events occurring.

Such methods are only valid for a large number of particles due to the probabilistic nature of quantum interactions; predicting the behaviour of a large number of neutrons is possible, tracking a single neutron is not.

Transport calculations solve what is called the neutron transport equation. This equation balances the number of neutrons in a system, accounting for losses and additions, and is based on the Boltzmann transport equation,

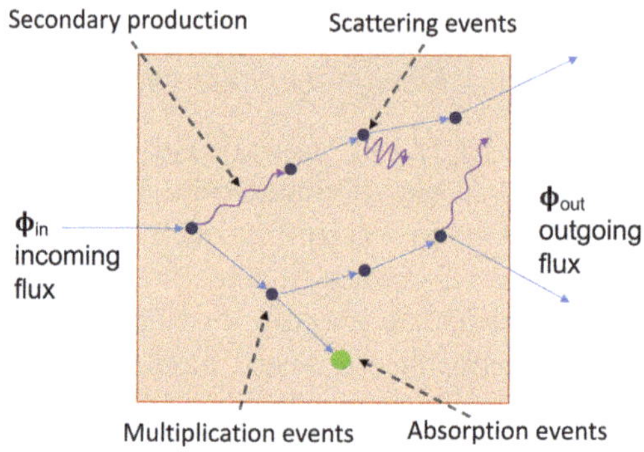

Figure 2.5. A cartoon showing how traversing material impacts the outgoing neutron flux spectrum.

$$\frac{1}{v}\frac{\partial \Phi}{\partial t} = -\underbrace{\Omega \cdot \nabla \Phi}_{1} - \underbrace{\Sigma_t \Phi}_{2} + \underbrace{\chi \int_E \mathrm{d}E \int_\Omega \mathrm{d}\Omega \nu \Sigma_f \Phi}_{3} + \underbrace{\int_E \mathrm{d}E \int_\Omega \mathrm{d}\Omega \Sigma_s \Phi}_{4}, \quad (2.4)$$

where the dependencies have been suppressed for simplicity. In equation (2.4), Φ is the neutron spectrum, t is time, v is the velocity, Ω is the solid angle and Σ_x is a macroscopic cross section ($=\sigma_x \rho_A$, here σ is a given cross section and ρ_A is the target atomic density).

The different numbered terms each represent different gains and losses of neutrons in a given volume:

1. Total neutron leakage out of the element.
2. Total losses and gains via interactions, e.g. absorption or multiplication.
3. Fission rate where χ defines the energy spectrum of produced neutrons, ν is the average number of neutrons produced per fission and Σ_f is the macroscopic fission cross section. Fission events are unlikely to occur in fusion reactor materials.
4. Scattering events which change the neutrons' energy and direction.

Solving equation (2.4) defines the neutron field in a given volume/element. In general, a reactor geometry will be divided into many such elements and this equation solved for each, with the flux tallied to account for neutrons from all sources. This can be solved via deterministic or Monte Carlo approaches. Commonly used transport codes used in fusion reactor development include MCNP [9] (developed at Los Alamos National Laboratory) and OPENMC [10].

2.2.2 Energy loss via scattering

The most common nuclear interaction which will occur is scattering. Scattering can be separated into two classes: *elastic* in which kinetic energy and momentum of the system as a whole is conserved and *inelastic* where some kinetic energy is lost to internal excitations of the target nucleus. It should be noted that neutrons do lose energy via elastic scattering as some kinetic energy is transferred to the recoiling nucleus—in fact, this is often the dominant energy loss mechanism. Materials which

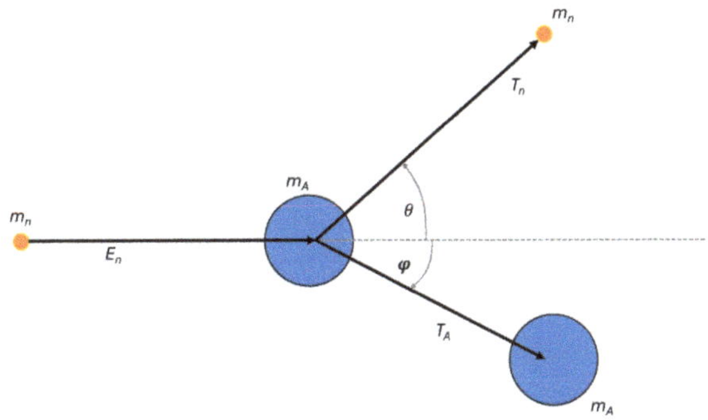

Figure 2.6. Scattering coordinates in the laboratory reference frame: m_n and m_A are the neutron and target masses, E is the kinetic energy before the collision, and T is the kinetic energy after the collision. θ and ϕ are the scattering angles.

shift the neutron spectrum from higher to lower energies are called *moderators* and the process is referred to as *moderating* the neutron spectrum.

How much energy is lost in a direct scattering event can be determined from conservation of energy and momentum. The coordinates of a direct scattering event are defined in figure 2.6. Under these definitions, the kinetic energy of the scattered neutron, T_n, in the laboratory reference frame can be obtained [11]:

$$T_n^{\frac{1}{2}} = \frac{(m_n^2 E_n)^{\frac{1}{2}} \cos\theta \pm \left(m_n^2 E_n \cos^2\theta + (m_A + m_n)[m_A Q + (m_A - m_n)E_n]\right)^{\frac{1}{2}}}{(m_A + m_n)}, \quad (2.5)$$

where m_n, m_A are the neutron and target masses, E_n is the neutron energy before the collision, θ is the scattering angle. Q is the reaction Q-value, where $Q = 0$ for elastic scattering. For the reaction $A(a, b)B$, the Q-value is found from the difference in the masses of initial and final reactants,

$$Q = (m_a + m_A) - (m_b + m_B + E_{\text{ex}}^{(B)}). \quad (2.6)$$

where $E_{\text{ex}}^{(B)}$ is the excitation energy of nucleus B. Equation (2.5) is double valued (due to the $\pm$, the expression is a solution to a quadratic equation) when $Q < 0$, this turning point defines the threshold energy for a given reaction, in the case of inelastic scattering this will be derived from the energy of the residual state.

Figure 2.7 shows the outgoing energies, as a function of incident energy, of neutrons elastic scattering off a number of different target nuclei, including ^{1}H, ^{2}H, ^{9}Be, ^{12}C, ^{56}Fe, ^{184}W and ^{208}Pb. The upper left panel shows that if there is no deviation in the neutron's path ($\theta = 0°$), the neutron's energy remains unchanged, whereas the other panels show cases where some energy is transferred to the target. The amount of energy transferred is dependent on the mass of the target. The neutrons lose very little energy while scattering off the heaviest targets (^{184}W and ^{208}Pb) and can lose almost all of it when scattering off the lowest mass targets. These observations make sense if we consider conservation of momentum: the closer in mass the target and neutron, the more linear momentum will be transferred in a

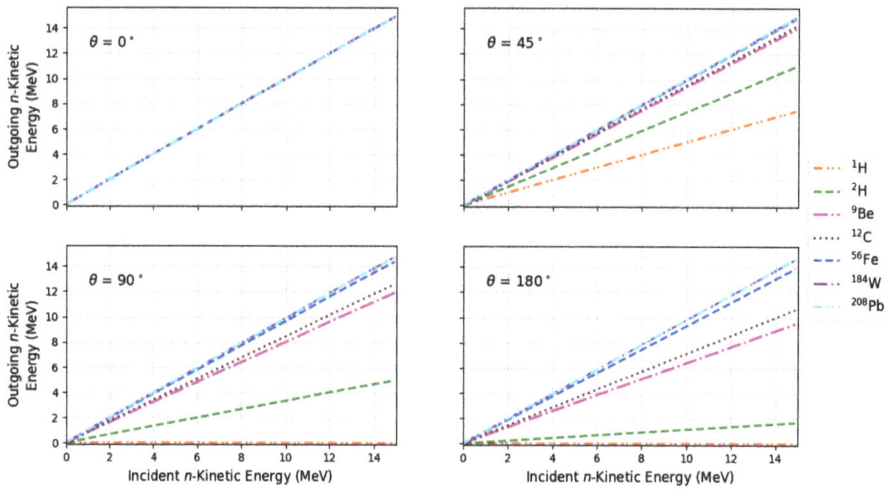

Figure 2.7. Plots of outgoing neutron energy against incident neutron energy for elastic scattering on a number of target nuclei. Results are presented for scattering angles of 0°, 45°, 90°, and 180° (see figure 2.6 for the coordinate system used).

collision. Scattering on ^{1}H provides some interesting results: above 90° almost all of the energy is transferred to the target (see the bottom panels of figure 2.7). The proton is the target most similar in mass to a neutron that one is likely to encounter in a fusion reactor. In fact, protons are slightly less massive, 938.272 MeV/c^2, compared to the neutron (939.565 MeV/c^2). This small mass difference means that the $(m_A - m_n)$ term in equation (2.5) is always negative, so above 90° (when $\cos\theta$ is also negative) almost all energy transfers to the proton.

The probability of collision will be determined by the density of a given material (how likely a neutron encounters a target) and the scattering cross section (how likely a scatter event occurs as opposed to another interaction). The angular emission will not be isotropic; the differential cross sections (cross sections differentiated with respect to a solid angle) will determine the physical distribution, with forward angles generally more favourable. The results presented in figure 2.7 are therefore representative but not sufficient to fully describe the evolution of a neutron spectrum. To do so, knowledge of all interactions and reactions and their impact on the neutron's energy and direction are required. Calculating such quantities *ab initio* would be extremely challenging and likely not fully accurate (nuclear reaction theory, at time of writing, is not a complete science) so an evaluated database is needed. Hence, a complete and accurate nuclear data library is essential in fusion reactor engineering.

2.2.3 Typical spectra and reactor components

The most notable feature of a fusion energy spectrum is the peak at 14.1 MeV—the energy at which neutrons are emitted from fusion reactions—which is much greater than that of typical fission spectra which peak around 2 MeV. A comparison between the expected neutron spectra from DT-fusion in ITER (magnetic

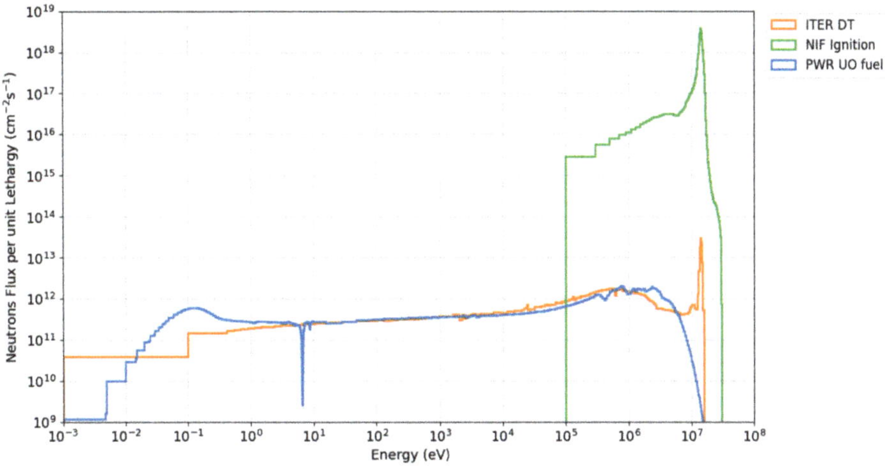

Figure 2.8. Neutron spectra from DT-fusion in ITER (magnetic confinement) and NIF (inertial confinement) compared with a typical fission spectrum (PWR). Both fusion sources show large 14.1 MeV peaks. It should be noted that the spectra do not appear smooth as a function of energy. This is an artefact of the spectra being tallied in energy bins (i.e. as histograms) which gives rise to the apparent step-like features. All spectra were taken from [12] and were obtained with MCNP [9].

confinement), and NIF (inertial confinement) and a typical pressurized water reactor (PWR; fission) is shown in figure 2.8. The 14.1 MeV peak is clear in the fusion cases and the higher low energy component in the fission spectrum. Figure 2.8 shows that different fusion devices can be quite different neutron sources; the NIF spectrum has such large magnitude as the units are given per second, NIF's fusion shots take place in a fraction of a second.

The neutrons emitted will impinge on the surrounding reactor structure; knowledge about how the neutrons will interact is provided by nuclear data. For reliable fusion simulations, nuclear data must be accurate across a wide range of energies and include reaction channels with high energy thresholds uncommon in fission environments.

Neutron spectra will evolve as the neutrons traverse the reactor geometry. The geometry will differ with reactor design, but fusion devices share common components and materials. As a neutron leaves a fusion plasma, it will encounter the following materials and interact with them as it traverses the reactor geometry:

1. **Armour and first wall.** This plasma-facing component acts to protect the other components from high temperatures and potential plasma discharges. Tungsten is often used due to its high melting point.
2. **Blanket.** The high intensity neutron spectrum is exploited for tritium breeding here. Many different materials are being explored for use in a blanket but the most common are advanced steel compositions.
3. **Vacuum vessel.** The structural component which contains the plasma chamber. Typically made from steel but will include water cooling.
4. **Bioshield.** The final shield beyond which the radiation dose should be zero. Usually concrete to shield against γ-radiation.

Knowledge of how the neutron field varies across a reactor is the starting point for all fusion nuclear analysis.

2.3 Tritium breeding

2.3.1 Introduction to tritium production

Breeding tritium is required for a DT-fusion reactor to have enough fuel available to operate. The region just behind the armour experiences high intensity neutron fields which can be exploited for tritium production. Tritium is produced via nuclear reactions so a material which when irradiated readily produces tritium and ideally no other radioactive products are required. There are many reaction channels which can produce tritium as such nuclear data libraries often give total tritium production cross sections[5]. These are the sum of reaction cross sections which produce tritium times by the multiplicity, the number of tritium nuclei produced by a given reaction.

It is well known that ^{6}Li has a high (n,t) reaction[6] cross section at low neutron energy (941 barns at thermal energies) making it an ideal candidate for tritium production target[7]. Figure 2.9 shows the total tritium production cross sections for

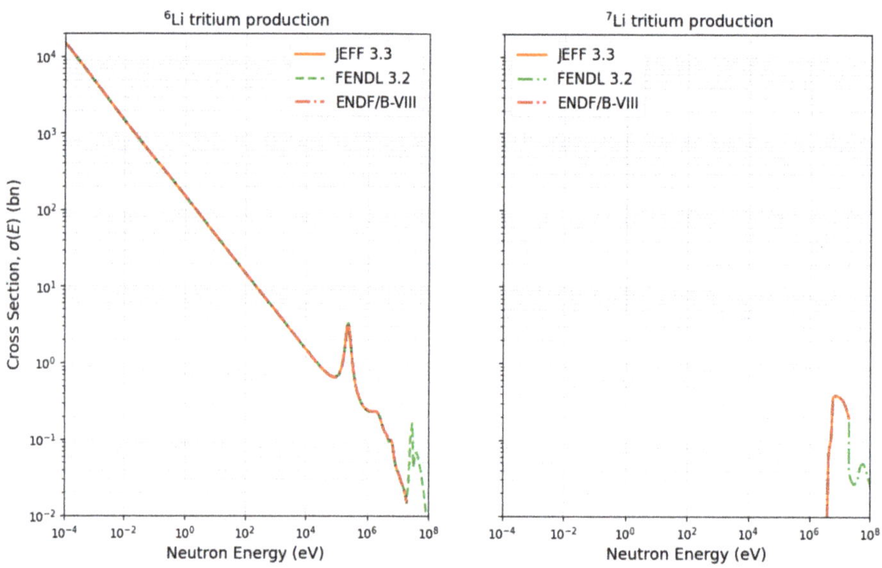

Figure 2.9. Total tritium production cross sections for ^{6}Li and ^{7}Li targets taken from the JEFF3.3 [13], FENDL3.2 [14] and ENDF/B-VIII [15] nuclear data libraries. The libraries show agreement with each other but the FENDL3.2 cross sections are evaluated to higher energies.

[5] The production of gases is often of interest in nuclear systems so total production of ^{1}H, ^{2}H, ^{3}He and ^{4}He are included alongside tritium.

[6] A neutron is incident and a tritium is ejected.

[7] It is worth commenting that this property of ^{6}Li has also led to it being incorporated into various classes of thermal neutron detector.

^{6}Li and ^{7}Li (naturally occurring Li isotopes) from a number of nuclear data libraries. The low energy component of the ^{6}Li cross section is very large, 3–4 orders of magnitude higher than the peak of the DT-fusion cross section in figure 2.1. The ^{7}Li cross section has a high energy threshold for tritium production, meaning that a pure ^{7}Li blanket could only make use of a limited region of the neutron spectrum.

Natural lithium only contains 7.59% ^{6}Li so lithium in a breeder blanket will likely need enriching, increasing the ^{6}Li proportion, to ensure enough tritium is produced. The ideal breeder blanket will produce at least a single tritium nucleus for each tritium used during fusion, i.e. the ratio of the tritium production rate in the blanket to the neutron production rate released from the plasma source should be $\geqslant 1$.[8] This ratio is called the tritium breeding ratio (TBR) given by

$$\mathrm{T\,BR} = \frac{N_t}{N_n} = \frac{\int_0^{E_{\max}} (N_{^7\mathrm{Li}}\sigma_{^7\mathrm{Li}}(E) + N_{^6\mathrm{Li}}\sigma_{^6\mathrm{Li}}(E))\phi(E)\mathrm{d}E}{\int_0^{E_{\max}} \phi(E)\mathrm{d}E}, \tag{2.7}$$

where N_x is the number density of species x, σ_y is the tritium production cross section for neutrons incident on nuclide y, $\phi(E)$ is the neutron flux as a function of energy (examples such as the neutron spectra shown in figure 2.8) and E is the neutron energy.

2.3.2 Designing a breeder blanket

A successful breeder blanket design will need to be tuned to the overall reactor design. There is unfortunately no universal breeder blanket and each reactor concept will have its own breeder blanket concept. However, there are many universal factors which need to be considered. This section will discuss these and show their potential impact.

2.3.2.1 ^{6}Li enrichment

As discussed, most breeder blanket concepts assume some level of ^{6}Li enrichment. As figure 2.9 shows, tritium is produced from ^{7}Li only from high energy neutrons so to make use of the complete energy spectrum a greater proportion of ^{6}Li will be required than that found in natural Li (which is 92.41% ^{7}Li). Figure 2.10 shows the tritium production reaction rate (derived from the ITER spectrum in figure 2.8 with the JEFF3.3 [13] cross sections) as a function of neutron energy. This is shown for various levels of ^{6}Li enrichment with the inset table detailing the total reaction rate. The results demonstrate that as the proportion of ^{6}Li increases, the reaction rate, and hence the amount of tritium produced, also increases. The only Li isotopic composition which has a lower reaction rate than natural Li is that of pure ^{7}Li, so it can be concluded that an increased proportion of ^{6}Li will be required for a sufficient TBR. It can be observed from figure 2.10 that a large enrichment fraction will be required[9] for an acceptable breeding blanket, therefore significant research will be

[8] An additional margin above 1 is needed for a self-sufficient reactor system to account for a range of losses, e.g. due to fundamental and technological aspects of the fuel cycle. See e.g. [16].

[9] Uranium fuel for fission reactors typically enriches ^{235}U content from 0.72% to approximately 4%.

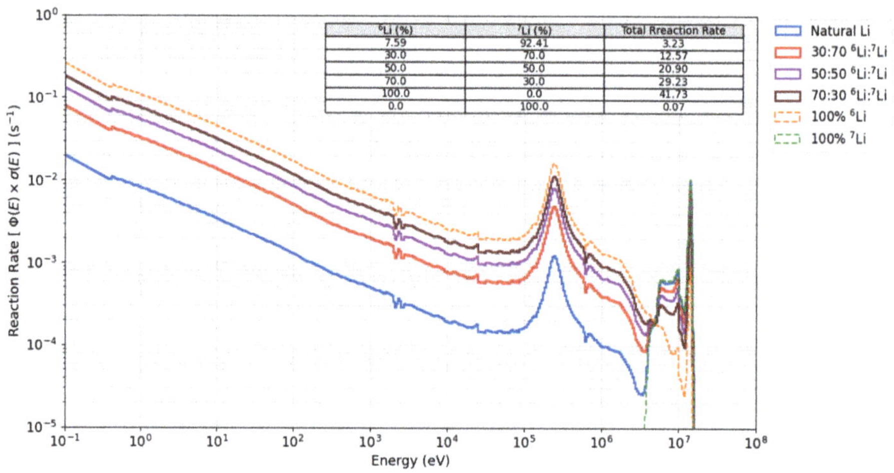

Figure 2.10. Tritium production reaction rate in Li (JEFF3.3 [13] cross sections in figure 2.9) as a function of energy for an ITER-like first wall spectrum (figure 2.8) for different ^{6}Li enrichment. The inset table shows the total reaction rate for each Li isotopic composition.

required to ensure that enough enriched Li can be produced to meet the needs of a global fusion industry.

Exactly what level of enrichment is best will depend on a number of factors and will require optimizing for each blanket design. One important factor is the exact neutron energy spectrum to which the Li will be exposed; if the spectrum is shifted to higher energies a greater proportion of ^{7}Li may be beneficial, if more lower energy neutrons are present, increasing ^{6}Li enrichment could improve TBR. The spectrum will change as the neutrons travel through the reactor, so the choice of structural and first wall materials will significantly impact tritium production.

2.3.2.2 Neutron multipliers

As the likelihood of tritium being produced from ^{6}Li (see figure 2.9) is greater for lower energy (below 10 MeV) neutrons, it can be useful to increase the number of neutrons with such energies present in the breeder blanket. As the neutrons traverse the reactor, they will often undergo inelastic scattering[10] and lose energy. Such reactions will effectively increase the number of lower energy neutrons. However, large amounts of material could be required and the energy transferred in neutron collisions could lead to local heating. Moreover, some fraction of neutrons will also be lost to competing capture and absorptive reaction processes. A possible alternative to this is to not rely on inelastic scattering but instead introduce materials which have large neutron multiplication reaction cross sections.

Neutron-induced neutron knock-out reactions will convert a single neutron to two or more lower energy neutrons (energy is required to liberate the additional

[10] Scattering events in which a neutron imparts energy to a target nucleus without transmuting it to different nucleus.

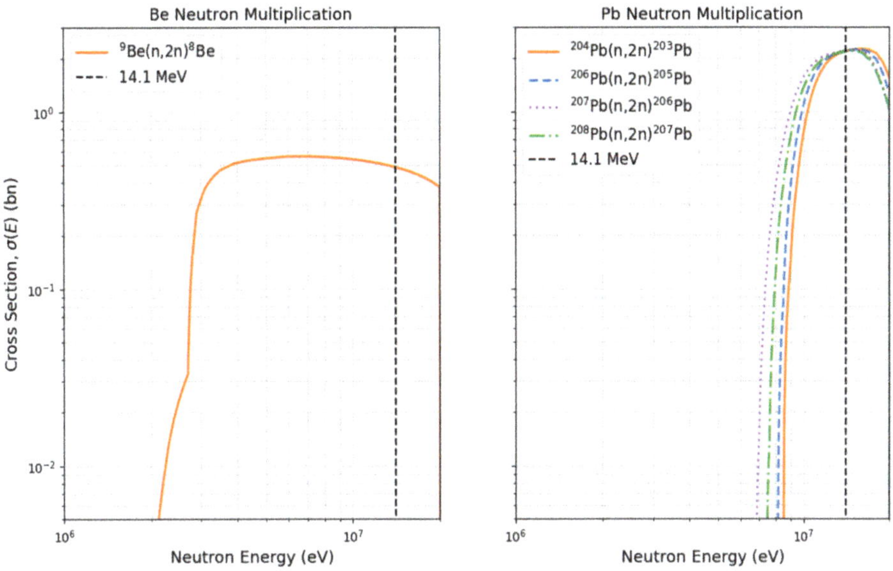

Figure 2.11. (n,2n) cross sections for stable Be and Pb isotopes taken from the FENDL3.2 [14] library. The vertical dashed line indicates 14.1 MeV—the energy of fusion neutrons.

neutrons) hence the term *multiplication*. Targets which have large (n,2n) and/or (n,3n) cross sections, particularly around 14 MeV would be the most suitable for use as potential neutron multipliers. Current research focusses on two candidates: beryllium and lead due to their large multiplication cross sections. Figure 2.11 shows the (n,2n) cross sections for stable Be and Pb isotopes taken from the FENDL3.2 [14] library. The Pb cross sections all peak around 14.1 MeV making them attractive for fusion, whereas the Be cross section allows multiplication to occur over a wider range of energies.

It should be noted that, under fusion-like neutron irradiation, other non-multiplication reactions will occur on Be or Pb targets. As such, multipliers need to be integrated into a blanket design such that they function as efficient multipliers. For example, let us imagine a hypothetical simple breeder blanket constructed in four layers: tungsten, an iron–water (90–10) mix, lithium (60% ^{6}Li) and a final layer of iron–water mix. Here, Tungsten is an approximation of the first wall armour and the iron–water mix that of cooled structural steel. Assuming that this blanket is exposed to a fusion neutron source such that the armour receives approximately 10^{14} neutrons cm^{-2} s^{-1}, the neutron spectrum which irradiates the Li will be calculated in three scenarios: the assumed blanket, with a 1 cm layer of Be before the Li, and with a 3 cm layer of Pb before the Li. Comparing the results, presented in figure 2.12, shows the impact of a neutron multiplier and allows the effectiveness of Be and Pb to be compared for this simplistic model.

The table in figure 2.12 shows that inclusion of the multiplier increases the total flux (neutrons s^{-1}) and the resulting tritium production reaction rate. Introducing a Pb layer increases the total flux by 3.5% with the Be layer giving a 5.2% increase.

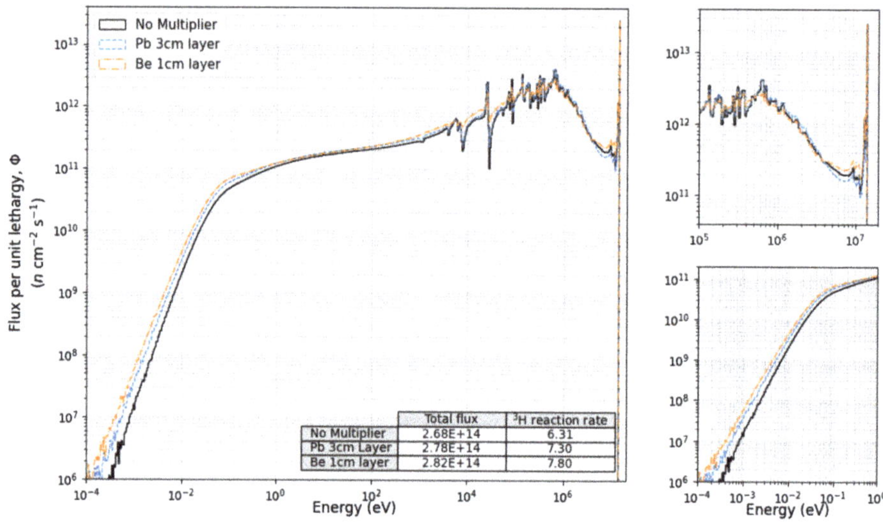

Figure 2.12. Neutron spectra found in a hypothetical breeder blanket with and without different potential neutron multiplier materials included. These calculations used the particle transport code OPENMC [10] and FENDL3.2 [14] nuclear data. The inset table shows total flux and tritium production reaction rate.

This may not seem significant but the tritium production reaction rate increases by 15.5% for Pb and 23.6% for Be. Be is the most efficient multiplier; a thinner layer produces a greater tritium production reaction rate in the Li.

While these results appear impressive, this is a very simple model intended to highlight the potential benefit of multipliers. The exact impact of altering the neutron spectrum on TBR is dependent on other aspects of the blanket such as Li enrichment, geometry and the surrounding materials. Not every breeder blanket design will require multipliers and relying on them to ensure sufficient TBR is often not desired. They are a useful tool to enhance an already fairly successful breeder blanket.

It should be noted that both Be and Pb are chemically toxic and can be challenging materials to handle in large quantities. After exposure to fusion neutrons, these will also contain radioactive nuclides (see section 2.5 which will cover these effects in more detail) which will hinder the handling and potential disposal of these materials. These factors influence breeder blanket designs away from multipliers where possible.

2.3.3 Current breeder blanket concepts

A successful breeding blanket not only needs to produce ample tritium, that tritium needs to be extractable and the materials able to survive the intense neutron fields. The impact of irradiation on material performance is discussed later (see section 2.5). The tritium produced will be needed to refuel the plasma during operation, without shutdown, and different blanket technologies are being explored to facilitate this. The most important consideration when engineering a breeding blanket is what form you breeder material takes; solid or liquid. It should be noted that at time of writing none of the breeding technologies exists outside of concept stage. The first functional breeding blankets may have very different designs.

2.3.3.1 Liquid breeders

At the currently expected operating temperatures of breeding blankets (greater than 500 °C), Li will be liquid (the Li melting point is 180 °C). Accordingly, a number of concepts related to flowing liquid breeders have been developed. A flowing liquid breeder has some advantages: the liquid can be piped through a tritium extraction system in cycle and the Li concentration can be maintained. There are disadvantages as well: the breeder will require heating at reactor start-up and when outside the neutron field (adding parasitic power load) and liquid metal corrosion may impact performance.

While Li's high specific heat capacity (3.58 kJ kg^{-1} K^{-1}) may allow it to serve as both breeder and primary coolant, pure Li is very difficult to engineer for. It is highly chemically reactive with H, N and O. In addition, Li metal is corrosive, meaning that designing blankets which will have sufficient lifetime is challenging. If neutron multiplication is desired then the multiplier material will need to be present alongside the flowing Li. Liquid breeder development has therefore looked into other potential breeders which have high Li concentrations but are easier to handle.

Many current liquid metal designs favour a LiPb eutectic. A eutectic mixture is one with a lower melting point than that of its constituent components. The exact melting point depends on the Li concentration but is considered to be approximately 235 °C (lower than the melting point of Pb which is 327.5 °C). As discussed in section 2.3.2, Pb is the leading contender as neutron multiplier so its inclusion should aid breeding performance. LiPb, by mass, is mostly Pb which is less chemically reactive than Li, lowering potential corrosion issues. LiPb will not be able to provide enough cooling, so a different primary coolant is required, usually water.

Non-metallic liquid breeders are also being investigated. Molten salts, such as FLiBe, are another candidate as a potential liquid breeder. Usually including Be (to act as a multiplier) this class of breeder typically has a higher melting point but greater heat capacity as compared to liquid metals.

2.3.3.2 Solid breeders

Li ceramics, which have melting points much higher than expected fusion reactor conditions, could be potential breeders. Li-orthosilicate, Li_4SiO_4, and Li-titanate, Li_2TiO_3 are leading candidates for solid breeder designs. Due to high operating temperatures, Be or Be-compounds are generally the preferred multipliers. These blankets are typically made up of a series of breeder–multiplier assemblies through which a coolant gas flows. They are designed such that any tritium produced will preferentially diffuse into the coolant for extraction. Helium is a common choice of coolant gas for solid breeder systems, but others (e.g. CO_2) have been considered.

2.4 Shielding

Given the highly intense neutron field produced by a fusion source, some reactor components will require protection in the form of shielding. While the external shields will be similar in concept to those found around fission reactors, the internal shielding will be needed to protect against high energy neutrons over a small volume and may make use of higher density concrete.

2.4.1 Nuclear reactions and shielding

What makes a material an effective shield will depend on precisely what is to be shielded against but all shields share a number of common features. In general, a material with a high density makes a good shield, this increases the likelihood of a neutron encountering a nucleus, as there are more atoms per unit volume. Shielding needs to stop, or at least diminish, a particle flux, so a shielding material will need to contain nuclei with high radiative capture and/or particle disappearance cross sections.

Radiative capture reactions are those with only photons outgoing, i.e. (n,γ) for neutrons. Disappearance reactions are those where the outgoing particles are different to those incident, e.g. (n,p), (n,d), (n,α). Either of these types of reactions effectively remove incident particles from the system, thus preventing or lessening the direct irradiation experienced by the material behind.

For incident neutrons, radiative capture reactions do not have a minimum energy; indeed, the cross sections are peaked at low incident energies—a consequence of slow neutrons being easy to capture. The emitted γ-rays arise from internal rearrangement to a lower energy configuration within the residual nucleus. Some reactions have a minimum energy at which a reaction can occur, this is defined as the incident energy where the reaction products are produced with zero kinetic energy. If the reaction Q-value is negative, a threshold energy for a reaction of the type $A(a, b)B$ can be defined

$$E_{\text{th}} = -Q\frac{m_B + m_b}{m_B + m_b - m_a}, \tag{2.8}$$

where m_z is the mass of the nucleus (or particle) z. Figure 2.13 shows binary reaction (single particle incident, single particle outgoing) cross sections for neutrons incident on ^{56}Fe. Given that a fusion device will be mostly steel, ^{56}Fe is likely to be the most common nucleus a neutron will encounter.

The reactions which emit particles all have $E_{\text{th}} > 1$ MeV, this is a consequence of the energy required to liberate massive particles. The radiative capture (n,γ) channel is open at all energies and has an especially large cross section at low energies. This is a common feature of all neutron capture cross sections. Materials containing nuclei with large capture cross sections are therefore a good shield for low energy neutrons, while materials containing nuclei with several channels open for threshold reactions can serve as a good shield for high energy neutrons. As discussed in section 2.2.2, neutrons will lose energy as they traverse a material via scattering events. As such, nuclei with large scattering cross sections can also make useful shielding materials as

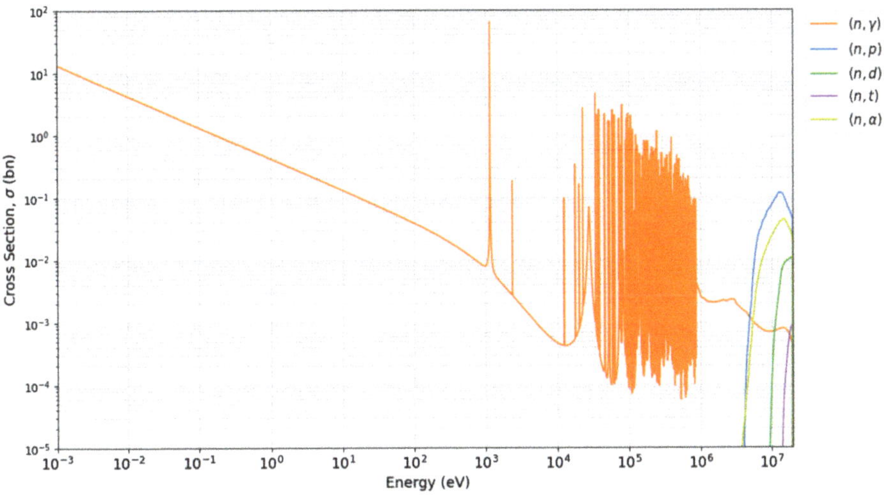

Figure 2.13. Binary reaction cross sections on ^{56}Fe. The (n,γ) reaction peaks at low energies; the heavy particle knock-out reactions all have higher energy thresholds. All data taken from JEFF3.3 [13].

neutrons can then lose sufficient energy to be readily captured. A downside to this approach is that a greater volume of material is needed and this might not be possible in-vessel.

Most shields are constructed of composite materials allowing for both direct removal of neutrons and evolution of the spectrum. Which is best in a given location will depend on the volume available, the environment (temperature, pressure etc) and the incident spectrum to be shielded against.

2.4.2 Neutron-shielding materials

Fusion devices will have shielding in many locations, for example, in-vessel shielding to protect sensitive components and/or diagnostics or the bioshield to prevent the neutron and photon fields irradiating workers. In-vessel shielding needs to reduce the flux of high energy neutrons over small distances, whereas shielding outside the vessel needs to prevent the leakage of lower energy neutrons and γ-rays.

To explore how different materials shield different energy neutrons a simple example can be used. The flux spectrum of neutrons passing through 10 cm layers of material, which are 10 cm from a neutron source, has been calculated. This has been performed for two isotropic 1×10^{19} ns^{-1} sources: one 14.1 MeV and the other 2 MeV. The materials studied are a mixture of elemental targets (Fe, W and Pb) and composite materials (water, concrete and borated polythene); the results are summarized in figure 2.14.

The total flux (see the middle panels of figure 2.14) is the results most directly relevant to shielding applications. For both the neutron energies considered, it can be seen that the flux generally reduces with depth into the material. This is true of all the different materials modelled. Notably, for several materials, the flux initially increases when transiting from vacuum into the initial layers of the material. This

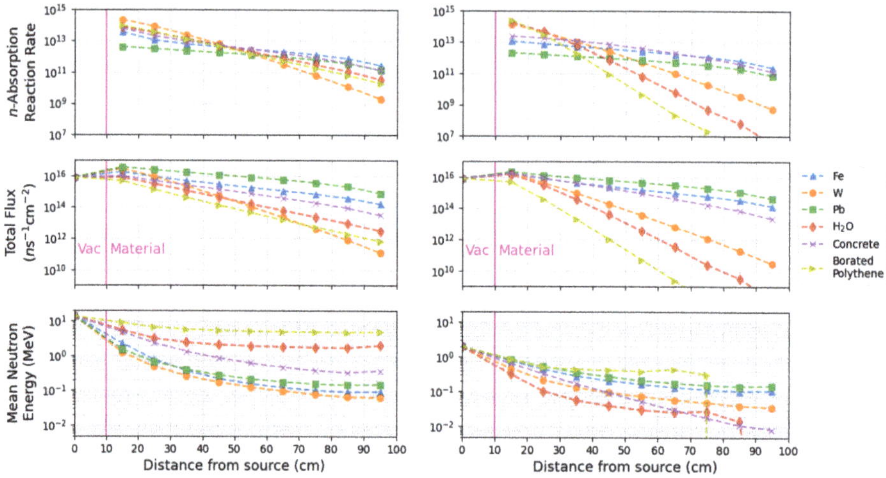

Figure 2.14. The absorption reaction rate (top panels), total flux (middle panels) and mean energy (bottom panels) for neutrons passing through various materials for an isotropic source 10 cm away. Results are shown for 14.1 MeV neutrons (left panels) and 2 MeV neutrons (right panels). The vertical magenta line indicates where the material begins.

counterintuitive effect is due to neutron multiplication reactions which are significant for some of the materials. It can be readily concluded from figure 2.14 that, irrespective of the initial neutron energy considered, the flux reduces most strongly with distance where the shielding material is water, borated polythene or tungsten (W). In some sense, it could be concluded that these materials are the *best* neutron shield, but that is not universally true and other factors need to be taken into account. For example, the resulting mean energy of 14.1 MeV neutrons (the bottom left panel of figure 2.14) passing through borated polythene or water remains greater than 1 MeV. If the objective were to shield higher energy neutrons, then these materials are clearly not the most efficient.

A few other remarks can be with regard to the shielding simulations discussed here. First, the absorption rate (see the top panels of figure 2.14) declines with distance, so the shielding is less efficient as the spectrum evolves. Second, shielding materials will inevitably to some extent undergo nuclear transmutation when absorbing the neutron dose. This can have serious consequences for material performance and reactor operations. This will be explored in the next section.

2.5 Impact of irradiation on reactor materials

As well as scattering, neutrons traversing reactor materials will induce nuclear transmutation via reactions on nuclei present. Transmutation refers to the conversion of one nucleus to another, this can both change elements and create radioactive isotopes, referred to as activation. Scattering events can also knock atoms out of alignment within a material changing its mechanical properties. How a material evolves during irradiation has important consequences for material selection, reactor maintenance and decommissioning planning.

2.5.1 Impact of nuclear transmutation

2.5.1.1 Introduction to transmutation and activation

The prediction and/or analysis of nuclear transmutation, via the capture and removal of nucleons, and possible activation, is an essential part of reactor design and operational planning. These calculations predict the evolution of the nuclide composition of a given sample under irradiation with a given spectrum. This evolution is dependent on the rate of creation and rate of decay of each nuclear species. The change in the number of a given nuclide, N_i, present in a sample is found by solving the ordinary differential equation,

$$\frac{\mathrm{d}N_i}{\mathrm{d}t} = \sum_j (\lambda_{j,\,i} + \sigma_{j,\,i}\Phi(t))N_{\,j}, \tag{2.9}$$

where N_j is the number of nuclei of species j, $\lambda_{j,i}$ is the decay constant for the transition from nuclide j to nuclide i, $\sigma_{j,i}$ is the cross section for reactions on j producing i, and $\Phi(t)$ is the time-dependent flux irradiating the sample. Nuclear data are required to solve equation (2.9); values for $\lambda_{j,i}$ will be taken from radioactive decay data libraries and $\sigma_{j,i}$ from reaction cross section libraries.

As can be seen from figures 2.1 and 2.8, reaction cross section and flux spectra are energy dependent, but equation (2.9) is not. $\Phi(t)$ in equation (2.9) is the total flux or energy integrated flux,

$$\Phi(t) = \int_0^{E_{\max}} \phi(E,\,t)\mathrm{d}E, \tag{2.10}$$

where ϕ is the neutron energy spectrum and $\sigma_{j,i}$ is the effective or collapsed cross section, given by

$$\sigma_{j,i} = \frac{\int_0^{E_{\max}} \sigma_{j,i}(E)\phi(E,\,t)\mathrm{d}E}{\Phi(t)}. \tag{2.11}$$

Thus the energy dependence of the reaction cross section is accounted for by weighting the contributions by the flux spectrum. If it is assumed that the evolution of the material does not change the flux spectrum, then the bracketed term in equation (2.9) is independent of the nuclide numbers N_j. This allows equation (2.9) to be solved for the complete set of nuclides, called the inventory, simultaneously,

$$\frac{\mathrm{d}\mathbf{N}}{\mathrm{d}t} = \mathscr{A}\mathbf{N}, \tag{2.12}$$

where $\mathbf{N}$ is a vector containing the numbers of each nuclide considered in the simulation and $\mathscr{A}$ is a $\mathbf{N} \times \mathbf{N}$ matrix which describes all possible nuclear transmutations. Solving equation (2.12) will give the evolved nuclide composition of a sample under irradiation. Many codes are available which can solve the inventory equations, such as FISPACT-II [17, 18] (developed at UKAEA, UK), ORIGEN [19, 20] (developed at Oak Ridge National Laboratory, US) and ACAB [21].

2.5.1.2 Radiological quantities and consequences of irradiation

From the evolved nuclide composition, a number of radiological quantities can be derived. Those most commonly desired from inventory simulations include:

- **Activity**. Nuclear disintegrations per second or becquerels (Bq). A measure of radioactivity given by

$$A = N\lambda, \tag{2.13}$$

where A is the activity, N the number of nuclei and λ the given nuclide's radioactive decay constant. An inventory's total activity is the sum of nuclide activities.

- **Decay heat.** Power emitted due to radioactive decay, usually given in kW. Nominally this includes contributions from all radioactive decay modes, but decay heating from specific emission (such as α particles) can be found:

$$D = A(\bar{E}_{\alpha} + \bar{E}_{\beta} + \bar{E}_{\gamma} + \cdots), \tag{2.14}$$

where D is the decay heat, A is the activity and $\bar{E}_x$ is the average decay energy for decay mode x. $\bar{E}_x$ is given in decay data libraries. The total decay heating from an inventory is the sum of nuclide decay heats (see later discussion of decay heat in fission reactors in chapter 3 and section 4.1.3).

- **Decay γ-spectrum**. The energy spectrum of emitted γ-rays from radioactive decay. The γ transitions for each nuclide are taken from a decay data library and weighted by the nuclide's activity in the inventory.

These quantities can be predicted with high accuracy if the supplied spectrum, material composition and nuclear data are accurate. An example of the results of an inventory simulation compared with experiment is shown in figure 2.15; this is a five minute neutron irradiation of the stainless steel, SS316. The results are taken from a set of decay heat benchmark experiments, covering a range of materials, performed using the Fusion Neutron Source (FNS) at the Japan Atomic Energy Agency (JAEA)[11] [22–24]. The left-hand panel of figure 2.15 shows a comparison between experimental data and calculations of the complete inventory decay heating using FISPACT-II in conjunction with either the TENDL2019, JEFF3.3 or ENDF/B-VIII nuclear data libraries. The right-hand panel of figure 2.15 shows the dominant nuclide activities and totals from the TENDL2019 results. The neutron spectrum used in these calculations was determined using MCNP [9].

2.5.1.3 Consequences of short-lived radioactivity in fusion reactors

Short-lived radionuclides will impact two important schedules of reactor operation: cooling and maintenance. After the neutron source has been shutdown, the reactor materials will still be heated due to the decay of short-lived isotopes. If this heating is sufficiently high, some reactor material could potentially begin to melt which would

[11] The compete data set and inputs for the inventory code FISPACT-II are available at https://nds.iaea.org/conderc/fusion/.

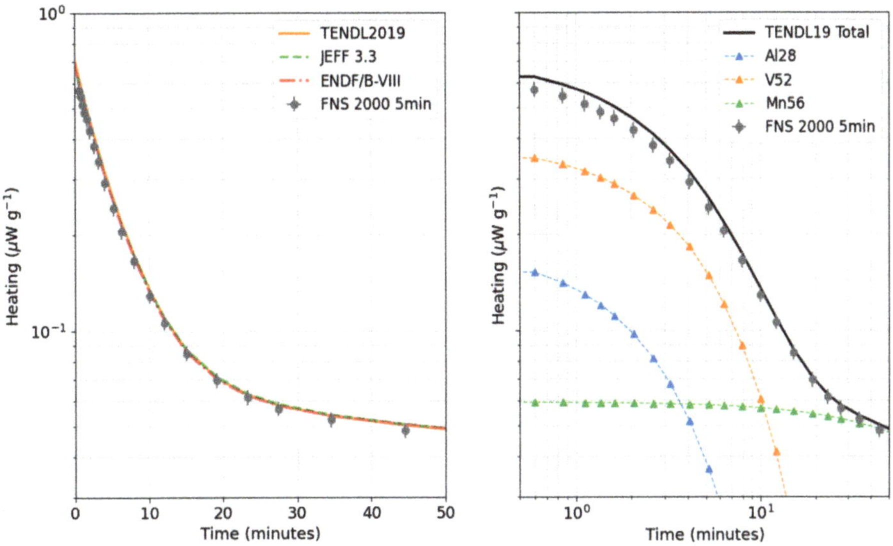

Figure 2.15. Decay heat measurement of SS316 steel after five minutes of neutron irradiation at FNS compared with simulations performed with FISPACT-II [17]. The left-hand panel shows the total decay calculated using the nuclear data libraries TENDL2019 [1], JEFF3.3 [13] and ENDF/B-VIII [15]. The right-hand panel shows the heating from nuclides which contribute at least 5% to the total for the TENDL2019 results.

be catastrophic for further operation. Using inventory simulations to predict the decay heating will help to define the post-shutdown cooling requirements of a fusion reactor.

In order to perform maintenance on a fusion device, access to the reactor will be required. Deciding whether this has to be performed remotely using robots or whether human-access can be allowed is dependent on the dose rate[12] which will be received. The dose is a consequence of the γ fields present after shutdown; each irradiated material in the entire reactor will contribute dose rate within the bioshield. The shutdown dose rate (SDDR) is calculated using photon transport to account for each component's contributions and position of measurement, with the γ sources determined from the inventory simulations. This dose rate will need to be sufficiently low before the bioshield or vessel could be breached and maintenance performed.

2.5.1.4 Consequences of long-lived radioactivity in fusion reactors

As well as resulting in the build up of short-lived isotopes, continued neutron irradiation will cause a number of long-lived isotopes to be produced. Due to their long half-lives, these will be present at reactor end-of-life and this residual radioactivity will need to be accounted for during decommissioning. If the activation is severe enough, reactor materials may require disposal as radioactive waste.

[12] This is the equivalent dose, given in sieverts (Sv). 1 Sv is the equivalent to 1 J of radiation energy deposited in 1 kg of human tissue.

Table 2.1. Hypothetical elemental weight percentages for material of in-vessel fusion reactor components.

Element	Composition 1	Composition 2
Fe	81.0	86.0
Cr	9.0	9.0
Ni	9.0	0.03
N	0.1	0.05
Mn	0.8	0.96
Nb	0.1	0.01
W	-	3.97

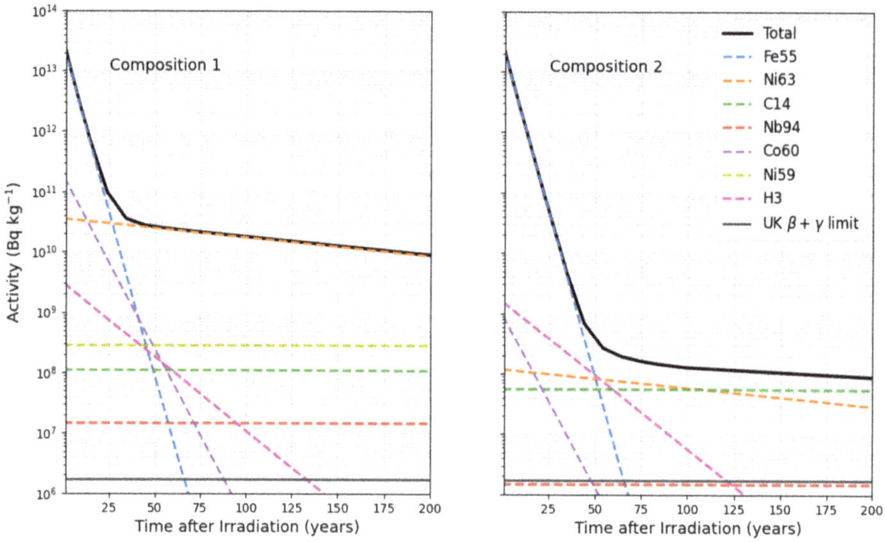

Figure 2.16. Activity of the material composition 1 and 2 given in table 2.1 after 1 year of ITER full power irradiation. A waste disposal limit on β and γ emitters is plotted for comparison.

Therefore, during materials selection and when planning reactor decommissioning, detailed inventory simulations using accurate neutron fluxes are needed to assess the waste burden.

For example, let us assume that the in-vessel components of ITER are constructed of materials with the elemental compositions shown in table 2.1. These will be exposed to one year of maximum power[13] irradiation using the ITER first wall neutron spectrum shown in figure 2.8.

The left-hand panel of figure 2.16 shows the activity of material composition 1 from one year to 200 years after the end of irradiation. Figure 2.16 also shows a limit on β

[13] In practice, this will be more than one year of real time since ITER operation would include downtime for maintenance.

and γ emitters of 1.2×10^7 Bq kg^{-1}; this can be used to determine if a given material can be characterized as low-level waste and potentially undergo near-surface disposal in the UK. It is expected that fusion power plant decommissioning will occur around 50–100 years after reactor shutdown.

Clearly, composition 1 produces a number of radionuclides which prevent it from being classed as low-level waste for over 200 years. A year of continuous irradiation from a fusion neutron source will result in activation but the levels seen in figure 2.16 would likely not be acceptable for a fusion power plant. The activation profile is dominated at early time steps by ^{55}Fe, which decays via electron capture and is created via a combination of the ^{54}Fe(n,γ)^{55}Fe and ^{56}Fe(n,2n)^{55}Fe reactions. ^{55}Fe has a half-life of 2.73 years; its activity will decay before 100 years, so it is a lesser concern from the perspective of radioactive waste.

Studying the contributing nuclides, and the pathways which created them, can help in exploring how changing the material composition might reduce some of the issues with activation. For example, ^{63}Ni and ^{59}Ni are produced via reactions on Ni isotopes, which can also contribute to the ^{60}Co population. Table 2.2 shows some possible reaction pathways from stable Ni isotopes to ^{63}Ni, ^{59}Ni and ^{60}Co.

Removing Ni from the composition would therefore lower overall activity by lowering the concentrations of these nuclides. However, a number of other long-lived nuclides (the flat lines in figure 2.16 due to the long half-life) are above the proposed limit. ^{94}Nb and ^{14}C would prevent low-level classification. These nuclides are produced via the ^{93}Nb(n,γ)^{94}Nb and ^{14}N(n,p)^{14}C reactions, respectively. Composition 2 listed in table 2.1 has reduced Ni and Nb content (introducing W as an alternative) and the results of the ITER-like irradiation are shown in the right-hand panel of figure 2.16.

Comparing the panels in figure 2.16 shows that reducing Ni and Nb leads to reduced long term activation: ^{94}Nb, ^{59}Ni and ^{60}Co activities have all been greatly reduced. ^{63}Ni activity is also reduced but this and ^{14}C would still contribute to the material not meeting the prescribed limit. This shows that when developing fusion reactor materials, the concentration of elements which can produce problematic long-lived radioisotopes need to be carefully controlled. This can prevent excessive

Table 2.2. Reaction pathways from Ni isotopes to the radioactive species ^{63}Ni, ^{59}Ni and ^{60}Co. Other reaction pathways will occur.

Nuclide	Pathway
^{63}Ni	^{64}Ni(n,2n)^{63}Ni
	^{62}Ni(n,γ)^{63}Ni
^{59}Ni	^{60}Ni(n,2n)^{59}Ni
	^{58}Ni(n,γ)^{59}Ni
^{60}Co	^{60}Ni(n,p)^{60}Co
	^{58}Ni(n,γ)^{59}Ni(n,p)^{59}Co(n,γ)^{60}Co

build up of waste critical nuclides and lower long-lived activation. Reducing impurities in reactor materials would also improve activation performance.

However, this alone would not remove all activated products. The amount of N is the same in both compositions in table 2.1 resulting in the same ^{14}C activities. N is added to steels to increase hardness so it will be unavoidable in a reactor constructed mostly of steels. Reactor operators will therefore need to take account of such activation in decommissioning and maintenance planning. This study would suggest that fusion reactors will produce radioactive waste from the components activated during irradiation and this has been confirmed in detailed studies of various materials in more mature reactor designs [25].

While fusion reactors will not produce equivalent waste streams to long-lived highly active spent fission fuel (see figure 5.1), they will produce activated structural materials and other contaminated wastes. How much waste is produced will depend on the size of the reactor, the exact materials and their neutron exposure, but to suggest that fusion reactors will not produce radioactive waste is incorrect. Only accurate nuclear data coupled with high fidelity neutron transport and activation calculations can provide reactor operators with the essential predictions needed for safe operation and decommissioning.

References

[1] Koning A J, Rochman D, Sublet J -C, Dzysiuk N, Fleming M and van der Marck S 2019 TENDL: complete nuclear data library for innovative nuclear science and technology *Nucl. Data Sheets* **155** 1–55

[2] Ruhlig A J 1938 Search for gamma-rays from the deuteron-deuteron reaction *Phys. Rev.* **54** 308

[3] Bretscher E and French A P 1949 Low energy cross section of the $D-T$ reaction and angular distribution of the alpha-particles emitted *Phys. Rev.* **75** 1154–60

[4] Chadwick M B, Paris M W, Hale G M, Lestone J P, Alhumaidi S, Wilhelmy J B and Gibson N A 2024 Early nuclear fusion cross-section advances 1934–1952 and comparison to today's ENDF data *Fusion Sci. Technol.* **80** S9–71

[5] Timofeyuk N K, Bailey G W and Gilbert M R 2024 Modelling the γ spectrum from *d–t* collisions *Phys. Rev.* C **110** 014612

[6] Tilley D R, Cheves C M, Godwin J L, Hale G M, Hofmann H M, Kelley J H, Sheu C G and Weller H R 2002 Energy levels of light nuclei $A = 5, 6, 7$ *Nucl. Phys.* A **708** 3–163

[7] Tilley D R, Cheves C M, Godwin J L, Hale G M, Hofmann H M, Kelley J H, Sheu C G and Weller H R 2017 *Energy Levels of Light Nuclei A=5* (Triangle Universities Nuclear Laboratory)

[8] Lawrence Livermore National Laboratory Achieves Fusion Ignition *Lawrence Livermore National Laboratory* https://www.llnl.gov/article/49306/lawrence-livermore-national-laboratory-achieves-fusion-ignition (Accessed: 8 Sept 2023)

[9] Aaron J *et al* 2022 *MCNP® code version 6.3.0 theory & user manual Technical Report* LA-UR-22-30006, Rev. 1 Los Alamos National Laboratory

[10] Romano P K, Horelik N E, Herman B R, Nelson A G, Forget B and Smith K 2015 OpenMC: a state-of-the-art Monte Carlo code for research and development *Ann. Nucl. Energy* **82** 90–7

[11] Krane K S and Halliday D 1987 *Introductory Nuclear Physics* (Wiley)

[12] Reference input spectra *FISPACT-II UKAEA* https://fispact.ukaea.uk/wiki/Reference_input_spectra (Accessed: 8 Sept. 2023)

[13] Plompen A J M *et al* 2020 The joint evaluated fission and fusion nuclear data library, JEFF-3.3 *J. Phys.* A **56** 181

[14] Forrest R A, Capote R, Otsuka N, Kawano T, Koning A J, Kunieda S, Sublet J -C and Watanabe Y 2012 *FENDL-3 Library Summary Documentation* IAEA

[15] Brown D A *et al* 2018 ENDF/B-VIII.0: the 8th major release of the nuclear reaction data library with CIELO-project cross sections, new standards and thermal scattering data *Nucl. Data Sheets* **148** 1–142

[16] Abdou M *et al* 2020 Physics and technology considerations for the deuterium–tritium fuel cycle and conditions for tritium fuel self sufficiency *Nucl. Fusion* **61** 013001

[17] Sublet J -C, Eastwood J W, Morgan J G, Gilbert M R, Fleming M and Arter W 2017 FISPACT-II: an advanced simulation system for activation, transmutation and material modelling *Nucl. Data Sheets* **139** 77–137

[18] FISPACT-II *UKAEA* https://fispact.ukaea.uk/ (Accessed: 13 Sept. 2023)

[19] Gauld I C, Radulescu G, Ilas G, Murphy B D, Williams M L and Wiarda D 2011 Isotopic depletion and decay methods and analysis capabilities in scale *Nucl. Technol.* **174** 169–95

[20] Origen *ORNL* https://www.ornl.gov/project/origen (Accessed: 13 Sept. 2023)

[21] Sanz J, Cabellos O and García-Herranz N 2008 ACAB Inventory code for nuclear applications *Technical Report* OECD-NEA

[22] Maekawa F *et al* 1998 *Compilation of benchmark results for fusion related nuclear data, Technical Report* JAERI-Data/Code 98-024 JAEA

[23] Maekawa F *et al* 1998 *Data collection of fusion neutronics benchmarking experiment conducted at FNS/JAERI, Technical Report* JAERI-Data/Code 98-021 JAEA

[24] Maekawa F, Wada M and Ikeda Y 1999 *Decay heat experiment and validation of calculation code systems for fusion reactor, Technical Report* JAERI 99-055 JAEA

[25] Bailey G W, Vilkhivskaya O V and Gilbert M R 2021 Waste expectations of fusion steels under current waste repository criteria *Nucl. Fusion* **61** 036010

IOP Publishing

Nuclear Data
Applications to society and industry
David G Jenkins

Chapter 3

Fission reactors and decay heat

Alejandro Algora, William Gelletly, Giuseppe Lorusso and Andrew Petts

It might, at first glance, be surprising that there are continuing nuclear data interests in the operation of fission reactors given the maturity of the underlying technology, the first operational reactor for power generation having been opened at Calder Hall in 1956. The situation is therefore rather different to that associated with fusion energy where the methodology and technology intended for future tokamaks is evolving rapidly (see chapter 2). While approved fission reactors operate safely and successfully and engineering models build in a tolerance to take account of the precision in available nuclear data, access to nuclear data with improved precision and accuracy may help to reduce the tolerances in engineering models and reduce uncertainty. Nuclear data aspects of fission reactors will be covered in the next three chapters. Here, we begin with a short historical introduction to the discovery of the fission process before moving on to an explanation of how nuclear reactors work. We will then turn to the topic of reactor decay heat and beta-delayed neutron emission which are critical for safe operation of nuclear reactors. Experimental efforts to improve data in both these areas will be described.

Concepts: Neutron-induced fission, fission reactor operation, neutron moderation, control rods, decay heat, beta-delayed neutrons

3.1 Historical introduction to neutron-induced fission

In the 1930s, Chadwick [1] had established the existence of a neutral particle—the neutron—with a mass similar to that of the proton. In addition, the Joliot-Curies [2, 3] had shown in 1933 that radioactive species could be created artificially. When bombarding aluminium with alpha particles they found that when the alpha source was removed the radiation from phosphorus nuclei could still be detected. They had observed the residual radiation from the $^{27}Al(\alpha,n)^{30}P$ reaction[1].

[1] It should be noted that this bland statement is correct but it took a great deal of thought and discussion before people realized it was so. The interested reader who wants to know more should read [4].

doi:10.1088/978-0-7503-5102-7ch3

© IOP Publishing Ltd 2025. All rights, including for text and data mining (TDM), artificial intelligence (AI) training, and similar technologies, are reserved.

It was realized that if one bombarded elements with neutrons one would produce nuclei decaying by β^- emission, thus producing residual nuclei with an atomic number greater by one. Fermi and his collaborators carried out many such experiments. Indeed, their original idea was to try to produce elements beyond U in this way. However, the results were not clearcut. They appeared to observe activity from Ra, which could be produced in the (n,2α) reaction on U, but the cross section seemed to be too large. Hahn and Strassmann [5] resolved the matter by using radiochemistry and showed that the activity was from the chemically similar Ba not Ra. It was then found that many other elements could also be produced in this process. Lise Meitner had been in close touch with this work even after having to flee Germany because of the persecution of Jews. She and her nephew Fritz, knowing that a lot of energy was released in this process, explained it [6] in terms of the U nuclei that are produced after neutron capture being unstable and dividing roughly in two. They called the process *fission*, a term borrowed from biology where cells divide in two. An important factor to understand the process was the use of the liquid drop model introduced earlier by George Gamow in 1929 [7].

As a typical example, figure 3.1 shows the idea of how neutron-induced fission of ^{235}U occurs. The neutron faces no Coulomb barrier and can be absorbed by the nucleus of ^{235}U. Since the nucleus is already deformed, the added neutron can deform it further and excite the resulting compound system to states where the splitting might be energetically favourable. It will start to separate into two roughly equal masses. Now the Coulomb repulsion between the two will overcome the nuclear force and they will separate into two nuclei, known as fragments. The fission fragments will vary in mass from about 70 to 160. Typically, they are neutron-rich and hence unstable. This means that a small number of neutrons, typically ~2.5 on average, will be emitted during the fission process. The fragments will then decay by

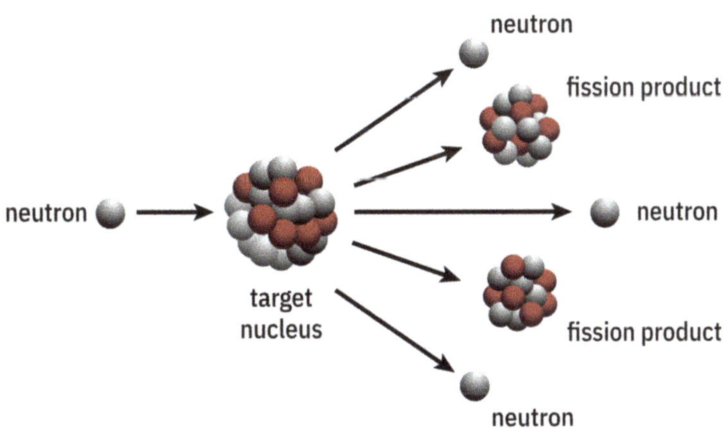

Figure 3.1. A simple view of the nuclear fission process. A neutron is absorbed by a target nucleus such as ^{235}U. As explained in the text, the uranium nucleus has a deformed shape and absorbing the neutron makes it further deformed. The Coulomb repulsion then causes it to separate into two fragments plus several neutrons. A typical example of the end products might be ^{144}Ba + ^{89}Kr + 3n.

a sequence of beta decays reaching nuclei on or near the line of stability. In some cases, one of the nuclei in the decay chain may be in a highly excited state and will decay by neutron emission rather than gamma emission. These delayed neutrons are important in terms of the operation of a nuclear reactor. As we shall see in section 3.4, they are also important in terms of the production of elements in stars.

Does all of this make sense in terms of energy? Earlier, in figure 1.2, which shows the binding energy per nucleon for stable nuclei as a function of the mass number A, we see that it is about 7.6 MeV per nucleon at ^{235}U. At half this mass, namely $A \sim 117$, it is $\sim$8.5 MeV per nucleon. So, the change in binding energy after fission is about 212 MeV. Most of this energy, about 90%, is emitted promptly in the form of the kinetic energy of the two fragments plus on average $\sim$2.5 neutrons with a mean energy of about 2 MeV. There remains another approximately 13% of the total energy released in the radioactive decay of the fission fragments.

Are there differences between the various transuranic nuclei when bombarded by neutrons? The answer is yes. In simple terms, one can imagine varying the combination of the two fragments inside the nuclear potential well and the height of the Coulomb barrier. For our purpose here, it is sufficient to notice that in figure 3.2, the cross section for fission is very large at low energies for ^{235}U. In the case of ^{235}U the cross section for fission falls roughly as $\frac{1}{v}$, where v is the velocity of the neutron. In contrast for ^{238}U, the cross section is zero until the energy is above $\sim$1 MeV. It turns out that this is very important, as we shall see below.

Putting some of the above facts together we see that, if we had a thick U target, the neutrons produced in fission can induce fission in other U nuclei. This can be repeated *ad infinitum* in a process called a *chain reaction*. This is characterized by k, the neutron multiplication factor, defined as the ratio of the number of neutrons in successive generations. Clearly if $k < 1$ then the process will die out. If $k = 1$ it will carry on at a steady rate, in principle, forever, but practically speaking until the uranium fuel runs out. If $k > 1$ a huge amount of energy will be released rapidly.

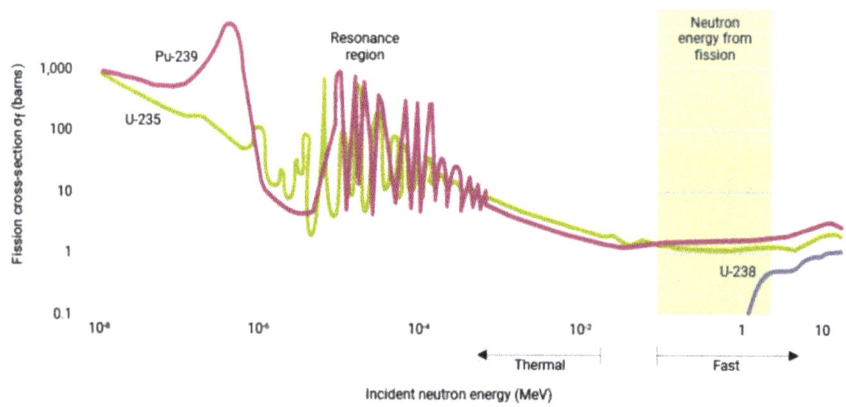

Figure 3.2. The figure shows the cross section for fission, on a logarithmic scale, for ^{235}U, ^{238}U and ^{239}Pu as a function of neutron energy in MeV. The shaded area indicates the energy region corresponding to neutrons emitted from fission reactions.

This last case is, of course, the basis for nuclear weapons. With $k = 1$ we have a steady source of energy and hence this has been utilized to produce power in the form of a nuclear reactor. It should be noted that we cannot create a reactor or a bomb with natural U, which consists of 0.7% ^{235}U and 99.3% ^{238}U. This is because of what happens to the neutrons. In general terms there are three possibilities: they can be scattered or captured by a nucleus or they can induce fission in another nucleus. As indicated above, ^{238}U will absorb too many neutrons with respect to the number of fissions and k will be less than 1. If, however, we produce uranium enriched in ^{235}U and can produce neutrons with sufficiently low energies then we will have a reactor producing a steady source of power.

^{235}U is not the only fissile nucleus but it is among the few that can be found in Nature. There are others such as ^{233}U, ^{239}Pu and ^{241}Pu that are not found in Nature, because they are relatively too short-lived, but they can be produced in nuclear reactions. In particular, they can be and are produced in reactors because of the very high neutron flux. If we start with ^{238}U and ^{232}Th as part of the reactor fuel, then we find that one can readily produce the first two. Thus

$$^{238}\mathrm{U} + \mathrm{n} \rightarrow {}^{239}\mathrm{U}\ (\beta^-/23.5\ \mathrm{min}) \rightarrow {}^{239}\mathrm{Np}\ (\beta^-/2.35\ \mathrm{d}) \rightarrow {}^{239}\mathrm{Pu}$$

$$^{232}\mathrm{Th} + \mathrm{n} \rightarrow {}^{233}\mathrm{Th}\ (\beta^-/22.3\ \mathrm{min}) \rightarrow {}^{233}\mathrm{Pa}\ (\beta^-/27.0\ \mathrm{d}) \rightarrow {}^{233}\mathrm{U}.$$

We see that if ^{238}U and ^{232}Th are present in the high neutron flux of a reactor, we can produce the fissile isotopes, ^{239}Pu and ^{233}U, following neutron capture plus two β^- decays. In other words, we can breed these nuclei and they can be used as fuel. In a reactor fuelled by uranium, ^{239}Pu is naturally produced because of the ^{238}U in the fuel, as is ^{241}Pu since it is produced by two neutron captures on the ^{239}Pu. Accordingly, as the reactor runs more and more of the energy is produced by the fission of the Pu isotopes, particularly the lighter of the two. Note that the ^{241}Pu is of no interest as a basic reactor fuel since it is too short-lived. There is interest in thermal reactors based on Th. They would have some advantages. For example, there is at least three times more Th in the Earth's crust than U and so one could fuel nuclear reactors for a much longer period in the future. They could also be used in the form of a liquid fuel and if there was a failure of the cooling system the reactor could be rapidly drained to make the reactor subcritical and thus stop reactions. This could be an important security measure for safer new reactors.

3.2 Nuclear reactors

Figure 3.3 shows the layout for a typical nuclear reactor power plant. It is a reactor based on fission induced by thermal neutrons. These are neutrons with, on average, about 0.025 eV energy where they are in thermal equilibrium with their surroundings. The reader will note that at this energy the cross section for fission of ^{235}U is very large (see figure 3.2).

The central core of a reactor usually consists of a moderating material, nuclear fuel and control rods. The disposition and material type of these three broad components differs depending upon reactor design. The core is usually organized with a set of fuel rods or assemblies surrounded by moderating material. The heat that is generated in

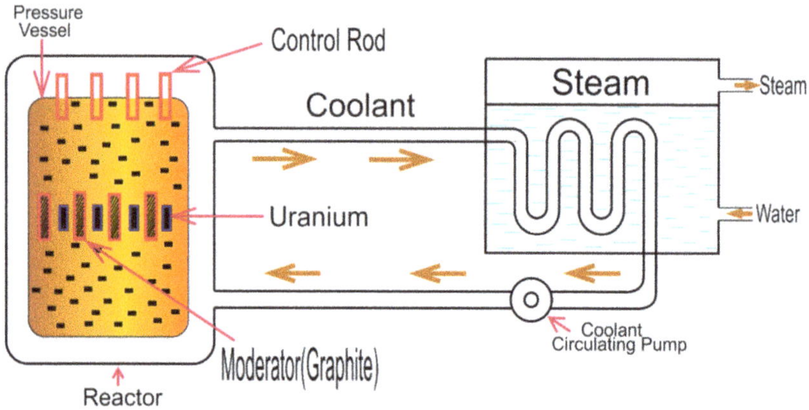

Figure 3.3. Typical layout of a fission reactor. There is a pressure vessel containing the U fuel surrounded by moderator material, here shown as graphite. The elastic scattering of neutrons in the moderator reduces their energies to the thermal value of 0.025 eV, where the ^{235}U fission cross section is 582 barns. There are also control rods, which can be inserted or withdrawn to control reactor operation. These rods contain materials such as B, Cd, Ag, Hf or In which readily absorb neutrons. The choice depends on the reactor design. The heat generated is removed by a circulating coolant to a heat exchanger where steam is generated that is used to drive turbines and produce electric power. Note that each reactor type will again use a different coolant. It may be light or heavy H_2O or a gas.

the core is extracted with a circulating fluid, referred to as the primary coolant. Control rods are arranged either within or between fuel assemblies to manage reactor power. A variety of different moderators have been used depending on reactor design. The key properties for a moderating material are a low neutron absorption cross section and low-Z to increase energy transfer per collision. In the UK, advanced gas-cooled reactors (AGRs) are graphite-moderated. The AGR design was itself an evolution of another UK-developed, graphite-moderated reactor—the MAGNOX reactor. Throughout the world the most common moderating material in use is light water —used in the prevalent pressurized water reactor (PWR) and boiling water reactor (BWR) reactor designs. The PWR and BWR designs use water as both moderator and primary coolant. It is important for reactor efficiency that the moderator/primary coolant has good heat transfer properties. The AGR and MAGNOX designs used a separate gas circulating through channels within the graphite core as a primary coolant. Both reactors used CO_2 as it was cheap and relatively inert with a low neutron capture cross section and had reasonable thermal properties. More exotic reactor designs such as CANDU use heavy water as a moderator and primary coolant. Heavy water has an advantage over light water due to its lower neutron absorption cross section. It is, however, much more expensive. Ironically, more advanced reactors in development are considering helium-cooled graphite-moderated cores, once considered for the AGR design. Helium has superior thermal properties to CO_2 but is more expensive. Helium-cooled reactors will be able to operate at much higher temperatures, improving thermal efficiency and unlocking the ability to provide high-temperature process steam used in heavy industry. The disposition of fuel within the reactor is reactor-design dependent. Typically, low-enrichment uranium in the form of ceramic pellets are contained within fuel pins made of either stainless steel or Zircaloy. Enrichments vary for different reactor designs, but are typically of the order of 3.5%. Fuel pin cladding material is chosen primarily for structural properties. It acts as the primary barrier to fission product release and must be able to tolerate high-temperature, high-radiation environments and maintain its structural integrity. The material will ideally have a low neutron absorption cross section to maximize fission yield. In the AGR reactor, fuel pins are grouped into fuel elements consisting of 36 pins arranged in three concentric rings. Eight of these elements are stacked into an assembly with around 300 assemblies loaded into the graphite core. Control rods are situated between four of these assemblies to manage reactivity. The typical arrangement for a PWR and BWR is different. Here, fuel pins are arranged in a 17×17 array to produce a fuel assembly. Within this array is a ‘grey’ rod used to help manage reactivity. The reactor core will consist of around 200 fuel assemblies. More exotic reactor designs such as some high-temperature gas-cooled reactors (HTGCRs) and molten salt reactors (MSRs) plan to use TRISO fuel. This fuel design is currently in development and encapsulates fuel kernels into three ceramic layers and promises to be safer and more efficient than current fuel designs.

The core is usually surrounded by a neutron reflector made of material with a high cross section for scattering neutrons. This is to minimize the number of neutrons lost from the system by scattering them back into the core, thus improving efficiency. Typically graphite or beryllium are used in common reactor designs. To manage reactor power, and ensure the ability to safely shutdown the reactor, control rods are deployed throughout the active core. Control rods usually are made of stainless steel doped with a material with a high neutron absorption cross section such as boron or cadmium. Standard reactor designs employ two types of control rod. ‘Black’ or ‘coarse’ rods have a higher concentration of boron or cadmium and are used for reactor shutdown only. These control rods will remain outside of the core when the reactor is at power and will fully insert when the reactor trips or is manually shutdown. ‘Grey’ or ‘trimming’ rods have a lower concentration of boron or cadmium and can be moved in or out of the core (usually by automatic control systems) to ensure that the neutron multiplication factor for the system stays close to 1 and to manage reactor power levels. The entire system is surrounded by a pre-stressed concrete pressure vessel (PCPV). This acts to absorb the intense gamma radiation and neutron flux generated and acts as the secondary barrier to fission product release should an accident occur.

Although reactor core design may vary, electricity generation at nuclear reactor sites is reasonably common in its approach. The primary coolant will be circulated through the reactor where it will remove heat from the core. The primary coolant will then circulate through the steam generators (sometimes referred to as boilers) and transfer its heat to the secondary coolant before being recirculated to the reactor. To avoid contamination of the ‘conventional’ plant, there is no direct contact between primary and secondary coolants. Within the boilers the secondary coolant (usually high purity demineralized water) is converted to steam which then drives a turbine and produces electricity which is exported to the grid system. Roughly speaking there are currently about 440 nuclear reactors in operation worldwide contributing power to national electricity networks. The reader should note that there are also some reactors used for research purposes, others for medical isotope production as well as power sources for submarines and ships. These designs vary, but, typically, they are PWR variants that employ uranium fuels more highly enriched than the fuel used in commercial power reactors. The reader should note that in order to accurately assess reactor conditions and states, knowledge of how radiation interacts with these materials is key. For this, accurate nuclear data are required. How these data and information are used in practice is discussed in detail in chapter 4.

3.3 Reactor decay heat

In a reactor based on ^{235}U, about 7% of the energy released is due to the beta, gamma and alpha radiation emitted in the decay of the fission products. The exact figure depends on the fuel that is used in a particular reactor. This is commonly called reactor decay heat (RDH). A reliable estimate of the energy that is released is important and, in a sense, it is obvious why. When the reactor is turned off, for whatever reason, energy is still released in the decay of the fission products that have been created. Accordingly, we must continue to maintain the reactor coolant. The decay heat decreases as a function of time after shutdown of the reactor since many different fission products, with widely varying half-lives, are produced. In normal operation, fuel is removed or exchanged at regular intervals. Spent fuel has to be stored. Accordingly, there is an elaborate waste management process with the fuel having to be removed, stored temporarily whilst the radioactivity decays to a suitable level, then transported in storage flasks and finally placed in long term storage. Although long term storage is not difficult in principle, it is a controversial matter because of public attitudes to nuclear radiation.

Failure to maintain cooling after shutdown can have catastrophic consequences such as what happened at the Fukushima power plant on the east coast of Japan. The reactor was shutdown because of an earthquake but the associated tsunami caused flooding in the emergency generator room, which meant that the emergency generators providing electrical power did not work. The result was a very serious accident.

Reactor decay heat has been a matter of keen interest from the birth of the nuclear industry. One can find an early attempt to calculate it in Way and Wigner [8]. Summation calculations are now the usual way of calculating decay heat. Here we can write

$$F(t) = \sum_i (E_{\beta,i} + E_{\gamma,i} + E_{\alpha,i})\lambda_i N_i(t), \tag{3.1}$$

where E_i is the mean decay energy of the ith nuclide, whether it be a charged particle (β or α) or a gamma-ray (γ), λ_i is the decay constant of the ith nuclide and $N_i(t)$ is the number of nuclides of type i at the cooling time t. Although E_α is included here it makes, of course, only a small contribution in a reactor at short cooling times. One can see that basically this equation describes the sum of the various decay activities, $\lambda_i N_i$, times the mean energies E_i released per decay to obtain the total power $F(t)$ at time t. Hence the name summation calculation. For practical reasons, since beta and alpha particles are much more readily absorbed than gamma-rays, their contributions are separated.

Clearly, if one is to employ equation (3.1), we must make use of a large amount of data since several hundred fission fragments are involved. The λ_i values are obtained from the half-lives where

$$\lambda = \frac{\ln 2}{T_{1/2}}. \tag{3.2}$$

It is obvious that if we have knowledge of the details of the decays of all the fission products produced then we can readily calculate how the decay heat varies with time (a similar approach for fusion reactors was presented in section 2.5.1). There are a number of international databases (e.g. JEFF, JENDL and ENDF) that contain data that have been evaluated. However, if the properties of the decaying nuclei are not known with precision, then such calculations would be of limited use. For example, if we wished to determine the mean energies released per decay in gamma or beta decay, we must know the distribution of beta feeding to the levels in the daughter nucleus in each individual decay.

It turns out that there is a particular problem here. First, how do we determine the beta feeding to a particular level? A simple example is given in figure 3.4. This shows a beta decay that feeds a single level in the daughter nucleus, i.e. feeding of 100%. This level then decays 100% to a lower level that then decays to the ground state. In any decay, the beta feeding is simply given by the difference in intensity of the electromagnetic transitions[2] feeding and de-exciting each level. In our simple example, the beta feeding to the two levels is then clearly 100% and 0%. What happens, however, if we do not observe the first gamma-ray? Now we would seem to

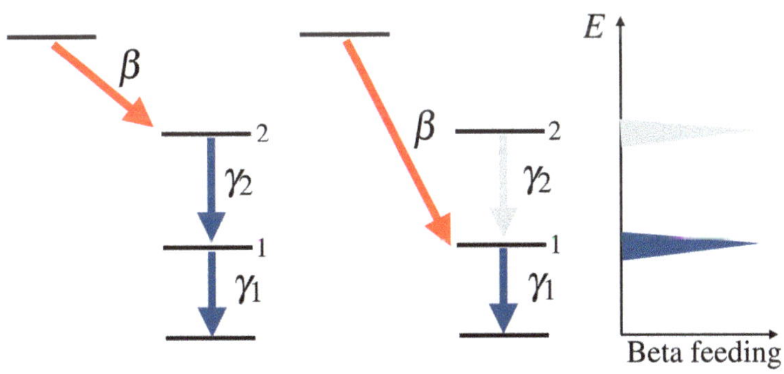

Figure 3.4. Simplified picture of a β decay where only one excited state is populated and it de-excites by the emission of a γ cascade, as shown in the left-hand panel. The central panel presents the pandemonium effect, in this example represented by missing, or not detecting the transition γ_2. The right-hand panel represents the displacement of the β-decay intensity because of the non-detection of the transition γ_2.

[2] In the figure and simplified example that is discussed, gamma-rays are represented, but one has to take into account the full electromagnetic de-excitation process, which may also comprise the electron conversion process.

have beta feedings of 0% and 100%. This is a very simple example of the pandemonium effect [9]. In more complex decays where there is feeding to higher energy levels, we may not observe gamma-rays because they are very weak or because the efficiency of detection is low at high transition energies. Accordingly, these data would not be fit for our purpose since the beta feedings will be wrong. Germanium detectors are widely used and they are extremely useful in determining level schemes due to their superb energy resolution. They do, however, suffer the drawback of having limited efficiency even in the form of large arrays of germanium detectors such as AGATA [10] or GRETINA [11], which represent the state-of-the-art of such technology. As a result, it turns out that many level schemes established using them, but not all, suffer from the pandemonium effect.

How do we solve this problem? The answer is by taking a different approach using a total absorption gamma spectrometer (TAGS). Now the gamma detector is essentially a calorimeter and ideally would have 100% gamma-ray efficiency. In this case, we determine the sum of all gamma-ray cascades de-exciting a level populated in the decay. Figure 3.5 shows a comparison [12] of the deduced beta strength mentioned earlier, which is based on the feeding beta intensity from the decay of the 2^- state in ^{150}Ho to levels in ^{150}Dy recorded on the one hand with the Ge cluster cube, an array of six Euroball cluster detectors, and on the other with the GSI TAGS [13]. Not surprisingly the two spectra follow the same pattern but about 50% of the beta strength is missing in the Ge spectrum—an excellent example of the pandemonium effect.

The analysis of the data from a TAGS experiment is not simple or straightforward. We will not describe it in detail. Instead, the reader can find the detail in a number of publications (see [14, 15] for some reviews). Ideally, a TAGS would be 100% efficient but, of course, there have to be a number of holes in the detector system. Since the radioactive source should be in the centre, a beam of the relevant radioactive nuclei must enter directly or be carried there on a tape on to which it has

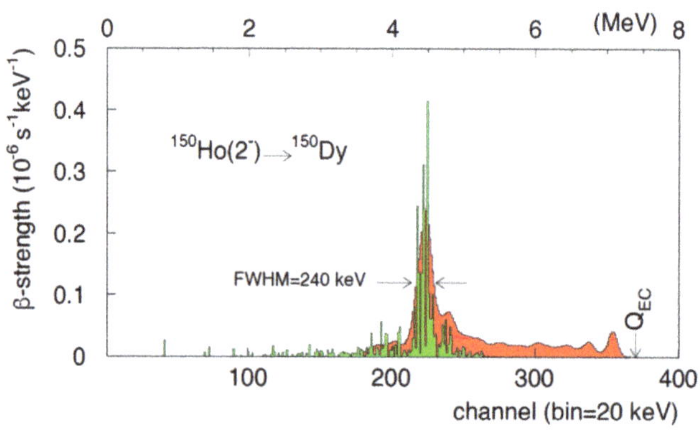

Figure 3.5. The beta strength spectrum deduced from the decay of the 2^- state in ^{150}Ho to levels in ^{150}Dy [12] measured with the Euroball cluster (green) and the GSI TAGS (red). One sees clearly that it is the same spectrum but 50% of the beta strength is missing in the Ge detector spectrum. See the text for details. (Reproduced with permission from [12]. Copyright 2003 the American Physical Society.)

been implanted. There must also, if required, be an opening if one wishes to use an ancillary detector for detecting some of the emitted charged particles. In addition, the detector must also be encapsulated. So, there are holes and dead material in the detector. This means that we cannot construct a 100% efficient detector.

As a result, to obtain the distribution of beta particles, one has to solve the inverse problem in the equation

$$d_i = \sum_j R_{ij}(B) f_j + C_i, \tag{3.3}$$

where d_i is the content of bin i in the measured TAGS spectrum, R_{ij} is the response matrix of the TAGS set-up and represents the probability that a decay that feeds level j in the level scheme of the daughter nucleus gives a count in bin i of the TAGS spectrum, f_j is the β feeding to the level j (our goal) and C_i is the contribution of the contaminants to bin i of the TAGS spectrum. The index j in the sum runs over the levels populated in the daughter nucleus in the β decay. How does one solve this inverse problem? A detailed description and explanation can be found in Algora *et al* [15]. In essence, one must first determine the response function $R(B)$. Then one can solve equation (3.3) using suitable algorithms. One very successful procedure was developed by Taín and Cano-Ott [16, 17]. They explored the use of several algorithms and found that the expectation maximization (EM) algorithm is the most useful since it has only positive solutions for the feeding distribution and no further assumptions are needed to solve the TAGS inverse problem. This procedure has been applied in a significant number of applications, see for example [15].

As indicated earlier, the fission process is complex with a very large number of fission products produced although not all produced in large amounts. In 2004, Yoshida *et al* [18] examined whether summation calculations were able to reproduce the experimental decay heat using the available databases at the time. He found that there was a poor description of the measurements in the 300–3000 s cooling interval. He also identified three beta decays that might be responsible for the deficiency and recommended that they should be measured using the total absorption technique. A collaboration between several institutions (Valencia–Surrey–IAEA) started discussions with experts in the field in order to identify additional cases that should be measured using the TAGS technique. This led to a committee of the International Atomic Energy Agency [19] which produced a list of decays that are significant contributors to reactor decay heat and should be studied with TAGS. Table 3.1 lists the 37 parent nuclides, identified by WPEC-25—the Nuclear Energy Agency working group—that should be measured using the total absorption technique to improve the predictions of the decay heat in reactors. These nuclides are of relevance for conventional reactors based on ^{235}U and ^{239}Pu fission. The criteria for placing a decay on the list are given in [19]. The committee went on to produce final lists for both U/Pu [20] and Th/U [21] based reactors.

Measurements using the TAGS method devoted to important fission decays started around 2004; the measurements being carried out at the IGISOL facility in Jyväskylä (Finland) [22, 23]. The reason that the measurements were carried out there is that many of the chemical elements concerned are refractory metals which have extremely

Table 3.1. List of parent nuclides identified by the WPEC-25 (Nuclear Energy Agency working group) that should be measured using the total absorption technique to improve the predictions of the decay heat in reactors. These nuclides are of relevance for conventional reactors based on ^{235}U and ^{239}Pu fission. Rel. (relevance) stands for the priority of the measurement. Nuclides marked with † are also relevant for $^{233}U/^{232}Th$ fuel, for which there are additional cases.

Isotope	Rel.	Isotope	Rel.	Isotope	Rel.
$^{86}Br^{\dagger}$	1	$^{99}Nb^{\dagger}$	1	$^{135}Te^{\dagger}$	2
$^{87}Br^{\dagger}$	1	$^{100}Nb^{\dagger}$	1	$^{136}I^{\dagger}$	1
$^{88}Br^{\dagger}$	1	$^{101}Nb^{\dagger}$	1	$^{136m}I^{\dagger}$	1
$^{89}Kr^{\dagger}$	1	$^{102}Nb^{\dagger}$	2	$^{137}I^{\dagger}$	1
$^{90}Kr^{\dagger}$	1	$^{103}Mo^{\dagger}$	1	$^{137}Xe^{\dagger}$	1
^{90m}Rb	2	^{105}Mo	1	$^{139}Xe^{\dagger}$	1
$^{92}Rb^{\dagger}$	2	$^{102}Tc^{\dagger}$	1	$^{140}Xe^{\dagger}$	1
^{89}Sr	2	$^{103}Tc^{\dagger}$	1	^{142}Cs	3
^{97}Sr	2	$^{104}Tc^{\dagger}$	1	^{145}Ba	2
$^{96}Y^{\dagger}$	2	^{105}Tc	1	^{143}La	2
$^{99}Zr^{\dagger}$	3	^{106}Tc	1	^{145}La	2
$^{100}Zr^{\dagger}$	2	^{107}Tc	2		

high melting points. The ion-guide technique, and more specifically the fission ion guide, allows the extraction of fission products independently of the chemical element. In this technique, fission is produced by bombarding a thin target of natural U with a proton beam. The fission products that fly out of the target are stopped in a gas and transported through a differential pumping system into the first accelerator stage of the mass separator. The dimensions of the ion guide and the pressure conditions are optimized in such a way that the process is fast enough for the ions to survive as singly charged ions. As a result, the system is chemically insensitive and very fast (sub-ms) [24] allowing the extraction of any element including those that are refractory. The set-up at Jyväskylä has another important advantage namely very high mass resolution. The beams are fed into the JYFLTRAP Penning trap [25], which was designed for high precision mass measurements. It can be used as a high-resolution mass separator for spectroscopy measurements with a mass resolving power of $M/\Delta M \sim 100000$, rather better than the equivalent resolving power of 500 of the separator magnet found in the beam-line. As a result of the high mass resolving power, errors related to contamination of the primary beam can be reduced greatly, since the measurements can be performed with *isotopically*, and in some cases even *isomerically*, pure beams.

Four experimental campaigns have been performed at the IGISOL facility to study the β decays that make important contributions to the decay heat and to the antineutrino spectrum in reactors using the TAGS technique [26–30]. One of the total absorption set-ups used in the experiments is presented in figure 3.6 (Rocinante TAGS). In a typical experiment, the pure radioactive beam obtained after the purification with the trap is implanted on a tape system, then the beta decay of the implanted ions is measured for a time interval, which is optimized according to

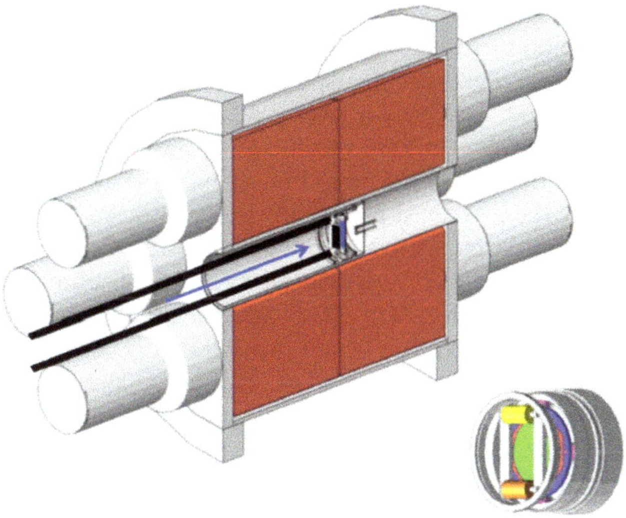

Figure 3.6. Schematic picture of half of the Rocinante total absorption spectrometer used in one of the experiments performed at the IGISOL facility of the University of Jyväskylä. The spectrometer is composed of 12 BaF_2 crystals. In the lower part, the endcap with the Si detector used as an ancillary detector is also presented (not to scale). The thick black lines represent the tape used to remove the remaining activity and the blue line represents the direction of the pure radioactive beam that is implanted in the centre of the spectrometer. (Reproduced with permission from [28]. Copyright 2017 the American Physical Society.)

the half-life of the decay and the half-lives of the descendants. These measuring cycles are repeated until enough statistics is collected for the measurements. In these experiments, ancillary detectors are used which allow one to select the events of interest. For example, in a TAGS measurement, coincidences with beta particles detected with a plastic or a silicon detector will dramatically reduce the impact of the ambient background on the measurements. We also need to measure well-known radioactive sources that allow us to calibrate the system and to evaluate how well the Monte Carlo models used characterize the system. This is relevant since the Monte Carlo model is used to calculate the response function of the system that is used in the analysis—see equation (3.3).

We need to recall that the ultimate goal of the measurements is to determine the feeding pattern of the decay free from the pandemonium effect. Once the feeding pattern is determined we can deduce the associated beta decay mean energies using the following relations:

$$E_\gamma = \sum I_j x E_j \tag{3.4}$$

$$E_\beta = \sum I_j x \langle E_\beta \rangle_j, \tag{3.5}$$

where E_j is the energy of the level j in the daughter nucleus, I_j is the probability of a beta transition to level j and $\langle E_\beta \rangle_j$ is the mean energy of the beta continuum populating level j.

The idea behind these approaches is relatively simple. In the case of the mean gamma energy, a good estimate can be obtained by summing the beta decay probability to each level multiplied by the energy of the level for all populated levels. This is easily understood since once a level is reached in the decay, it will probably decay by a gamma cascade, emitting the energy of the excited level in the form of gamma radiation. In the case of the mean beta energy, the approach is based on the sum of the product of the beta decay probability times the mean beta energy associated with the decay to each level. This is more complex than the mean gamma energy calculation, since each populated level determines first the maximum beta energy (or end point) for the beta transition, (Q_β—level energy), and for each level appearing in the sum we need to calculate a mean beta transition energy corresponding to all beta transitions with the corresponding end point. This last calculation requires that we make assumptions about the shapes of the beta transitions.

In table 3.2, we show some examples of deduced mean energies from TAGS data compared with the values determined from experiments using Ge detectors. It

Table 3.2. Mean gamma and beta energies (in keV) deduced from the studies of beta decays at Jyväskylä compared with the values deduced from high-resolution measurements. The highest energy level identified in each decay using Ge detectors and the Q_β values are also given for each decay. It should be noted that data from a decay with a large Q_β value measured with Ge detectors does not always mean that the case is afflicted with pandemonium. Decay cases with strong ground state to ground state feeding will suffer less from this systematic error. Details are given in the text.

Isotope	High. lev.	Q_β	$\bar{E}_\gamma^{HR}$	$\bar{E}_\gamma^{TAGS}$	$\bar{E}_\beta^{HR}$	$\bar{E}_\beta^{TAGS}$
^{86}Br	6768	7633(3)	3360(110)	3782(116)	1900(300)	1687(60)
^{87}Br	5793	6818(3)	3100(40)	3938^{+40}_{-67}	1660(80)	1170^{+32}_{-19}
^{88}Br	6999	8975(4)	2920(50)	4609^{+78}_{-67}	2240(240)	1706^{+32}_{-38}
^{91}Rb	4793	5907(9)	2270(40)	2669(95)	1580(190)	1389(44)
^{92}Rb	7363	8095(6)	170(9)	461(14)	3640(30)	3498(105)
^{94}Rb	6064	10 281(8)	1750(50)	4063^{+62}_{-66}	2020(90)	2450^{+32}_{-30}
^{95}Rb	4661	9284(21)	2050(40)	3110^{+17}_{-38}	2320(110)	2573^{+18}_{-8}
^{100}Nb	3130	6384(21)	710(40)	959(318)	2540(210)	2414(154)
^{100m}Nb	3647	6698(31)	2210(60)	2763(27)	2000(200)	1706(13)
^{101}Nb	1099	4569(18)	270(22)	445(279)	1800(300)	1797(133)
^{102}Nb	2480	7210(40)	2090(100)	2764(57)	2280(170)	1948(27)
^{102m}Nb	1245	7304(40)		1023(170)		2829(82)
^{105}Mo	2766	4953(35)	551(24)	2407(93)	1900(120)	1049(44)
^{102}Tc	2909	4532(9)	81(4)	106(23)	1945(16)	1935(11)
^{104}Tc	4268	5600(50)	1890(30)	3229(24)	1590(70)	931(10)
^{105}Tc	2403	3644(35)	671(19)	1825(174)	1310(210)	764(81)
^{106}Tc	3930	6547(11)	2190(50)	3132(70)	1900(70)	1457(30)
^{107}Tc	2680	4820(90)	511(11)	1822(450)	1890(240)	1263(212)
^{137}I	5170	5880(30)	1135(20)	1220^{+121}_{-74}	1897(15)	1934^{+35}_{-56}

should be emphasized that data from a decay with a large Q_β value measured with Ge detectors does not always mean that the case is afflicted with pandemonium. Decays with strong ground state to ground state feeding will suffer less from this systematic error (see, for example, the ^{101}Nb decay).

In figure 3.7, we see the impact of the TAGS measurements of thirteen decays [26–30] on the gamma and beta components of the reactor decay heat. The figure shows the decay heat data compilation made by Tobias [31] and the measurements by Dickens *et al* [32, 33] for ^{239}Pu and ^{235}U. In both cases, the gamma and beta components are shown separately. They are compared with the calculated values using the JEFF 3.1.1 database in which no new TAGS data have been included. They are also compared with results from the same database but now modified by the inclusion of the total absorption data for the mean energies. It is clear from the figure that the TAGS measurements have improved the calculations. It also shows that the TAGS measurements are essential if the summation calculations are to provide a proper assessment of the decay heat.

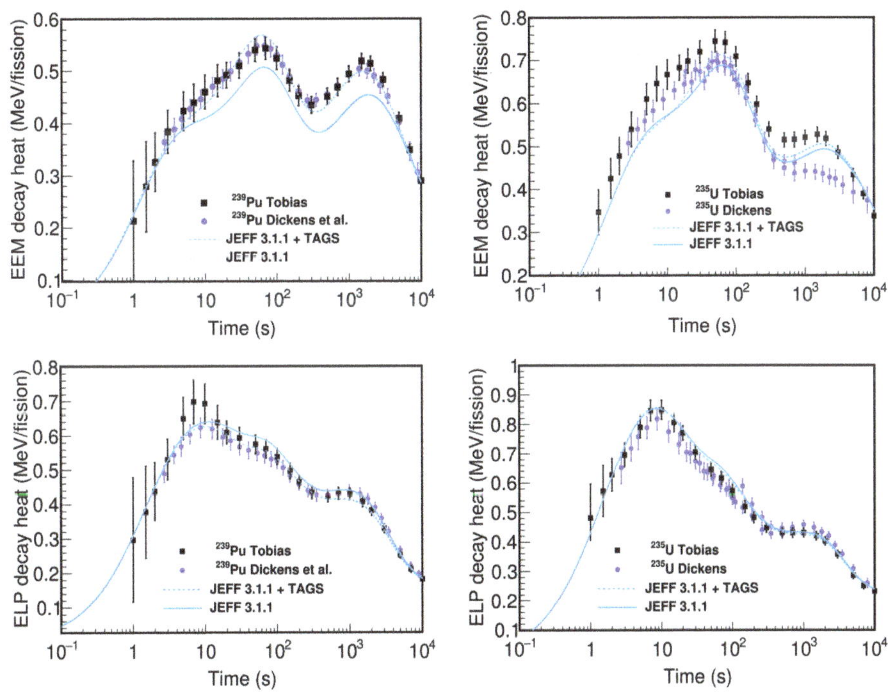

Figure 3.7. Impact of the inclusion of the total absorption measurements performed for thirteen decays ($^{86,\,87,\,88}$Br, $^{91,\,92,\,94}$Rb, ^{101}Nb, ^{105}Mo, $^{102,\,104,\,105,\,106,\,107}$Tc) [26–30] on the gamma and beta components of the decay heat calculations for ^{239}Pu and ^{235}U. Top left: gamma component of the ^{239}Pu decay heat. Top right: gamma component of the ^{235}U decay heat. Bottom left: beta component of ^{239}Pu. Bottom right: beta component of ^{235}U. (Reproduced from [15]. CC-BY-4.)

3.4 Beta-delayed neutrons

We now turn to β-delayed neutrons, which play an important role in nuclear structure, nuclear astrophysics and in the control of fission reactors. Let us first explain the phenomenon of beta-delayed neutrons before going on to discuss the role it plays in each of these three areas.

Earlier when we discussed how nuclear fission reactors operate, we effectively assumed that all of the neutrons emitted in the fission process were emitted promptly, i.e. within $\sim 10^{-14}$ s. However, in reality, it turns out that a small fraction, $\sim$ 0.68% in the case of ^{235}U, have a delayed emission. As we shall see below, this has important consequences and applications.

Delayed neutrons were first discovered in 1939 by Roberts, Meyer and Wang [34] who found that when uranium and thorium were bombarded by neutrons, some neutrons were detected as long as 1 1/2 min after the end of the bombardment. The decay period of these neutrons was measured as 12.5 $\pm$ 3 s. More precise characterization of the neutron decay period followed, and Booth, Dunning and Slack [35] also estimated that in the fission of a uranium atom, one delayed neutron was produced every 60 fissions.

Once it was experimentally demonstrated that the time it takes for fission is less than 10^{-12} s, and that photonuclear reaction induced gamma de-excitation occurs within 10^{-13} to 10^{-15} s, an interpretation of the phenomenon of delayed neutrons came from theory. Bohr and Wheeler [36] attributed the delayed neutrons to the beta decay of a fission fragment to levels of high enough energy in the daughter nucleus that they are above the neutron separation energy, S_n, and can thus emit a neutron. Other possible sources of such neutrons had already been ruled out by this time [37, 38]. It should be noted that levels populated above S_n do not necessarily solely emit neutrons since they may still emit gamma-rays in competition. The fraction of such beta decays that lead to neutron emission—the beta-delayed neutron emission probability P_n—is an important quantity wherever the process influences application of the fission process or our understanding of nuclear structure or astrophysics.

The process is illustrated and can be readily understood in the case of the decay of ^{137}I [39] shown in figure 3.8, where we see the beta decay to levels both below and above S_n in the daughter nucleus, ^{137}Xe. The levels below S_n decay exclusively by gamma-ray emission or internal conversion to the ground state of ^{137}Xe. Above S_n, the levels can decay either by neutron emission to ^{136}Xe or by gamma-ray emission to the ground state of ^{137}Xe. It should be noted that this is an unusual example in that the neutron decay channels all go directly to the ground state of ^{136}Xe.

As we move away from the line of nuclear stability, Q_β values rise and we find more and more beta-delayed neutron emitters. Eventually we reach nuclei where the beta decay Q_β value is sufficiently large that beta decay can populate levels above S_{2n}, the two-neutron binding energy and these levels can decay by emitting two neutrons simultaneously. In some cases, there may even be higher numbers of neutrons emitted, e.g. 3n. We are still a long way from measuring all possible beta-delayed neutron decays. In 2019, 326 such decays were known [40] (figure 3.9) but that number has grown and will steadily grow further as more high energy

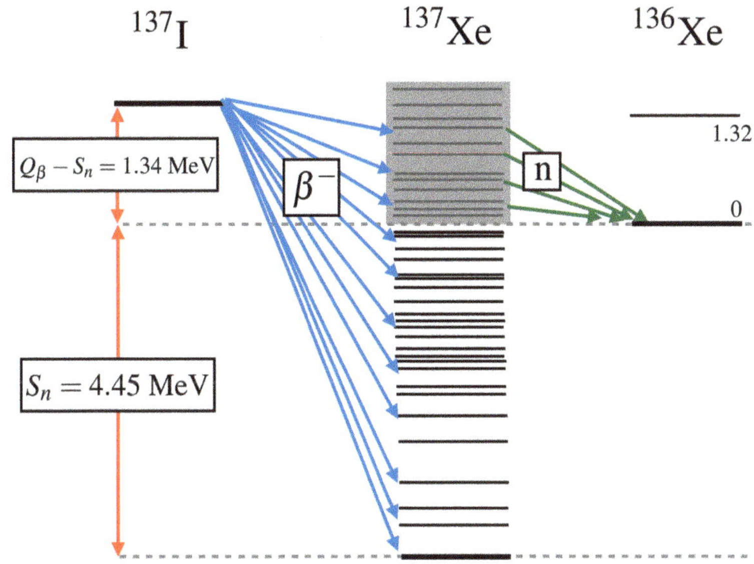

Figure 3.8. The beta decay of ^{137}I which proceeds to a large set of excited states in the daughter nucleus. Those states above S_n can decay either by neutron decay or by electromagnetic transitions to the ground state. The fraction of decays which do result in neutron emission is known as the beta-delayed neutron probability and denoted by P_n. It should be noted that no excited states are populated in the final nucleus ^{136}Xe which is relatively unusual.

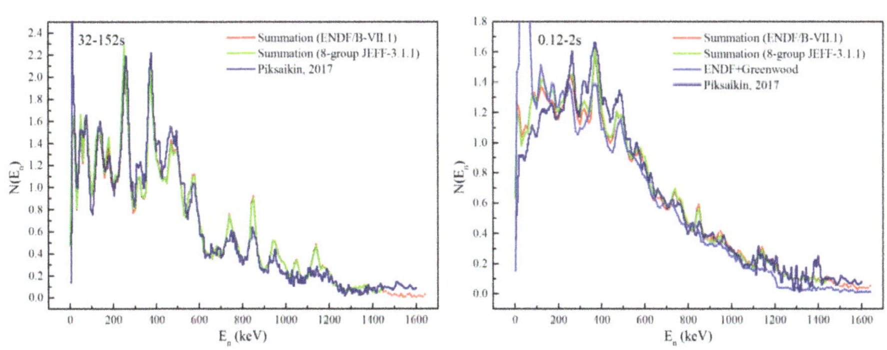

Figure 3.9. Energy distribution of neutrons emitted in the thermal neutron fission in ^{235}U. The measured spectrum (Piksaikin 2017) is compared with summation calculations using different databases and the 8-group approach. For more details see [40]. (Reproduced with permission from [40]. Copyright 2021 Elsevier.)

radioactive beams of high intensity are generated by new accelerators. Earlier we said that the delayed neutrons had a half-life of 12.5(30) s but the early measurements reflected a mixture of decays. In reality, we find each emitter has a different half-life.

For practical analysis and calculations, it was found that the decay of several precursors with similar half-lives can be approximated by a single exponential with an average half-life, called 'a group'. This helps to simplify the treatment of hundreds of beta-delayed neutron emitters. Neutron emitters of similar characteristics can be grouped and characterized by a common half-life, common fraction of the total number of neutrons, yield per fission, average energy, etc. The best fit to the parameters of the group can be found by statistical analysis. Increasing the number of groups improves the fit at the cost of more variables. Traditionally, depending on the fissioning isotopes, 4-, 5-, 6-group models have been considered satisfactory with group yields and average group half-lives.

There is also an 8-group approach, which is a little different and is based on the fact that more than 80% of the delayed neutrons are produced by approximately a dozen nuclei common to all fissioning species. The method increases the number of delayed neutron groups and fixes the half-lives in the new model to some suitable set of values that cover the known full range of precursor half-lives. Table 3.3 summarizes the basic information available for the eight delayed neutron groups.

The main advantages of this model consist in simplifying complex systems calculations in the case where more fissile isotopes are present. This is because the half-lives of the 8 groups are common to all the fissile isotopes in the group so the number of equations needed to describe the total delayed neutron activity is reduced from 30 to 8. A second benefit is the energy spectrum. The new model, using a consistent set of half-lives, results in more consistent sets of energy spectra. In general, the eight-group model has been found to yield better results.

Table 3.3. Eight-group beta-delayed neutron model details.

Group	Precursor	Group $T_{1/2}$ (s)	Main precursor(s)	Group average $T_{1/2}$ (s)
1	^{87}Br	55.6	^{87}Br	55.6
2	^{137}I	24.5	^{137}I	24.5
3	^{88}Br	16.3	^{88}Br ^{138}I	16.3
4	^{89}Br	4.35	^{93}Rb ^{89}Br ^{94}Rb	5.21
5	^{90}Br	1.91	^{139}I ^{85}As ^{98m}Y ^{93}Kr	2.37
6	^{98}Y	0.548	^{144}Cs ^{140}I	1.04
7	^{95}Rb	0.378	^{91}Br ^{95}Rb	0.424
8	^{96}Rb	0.203	^{96}Rb ^{97}Rb	0.195

The time distribution of delayed neutrons and the total delayed neutron emission per fission can be calculated once the P_n values, the fission product activities, fission yields and other nuclear data are known. For example, the number of neutrons emitted as a function of time is given in general by

$$n_{\mathrm{emit}}(t) = \sum_i P_{ni} \lambda_i N_i(t), \tag{3.6}$$

where P_{ni} is the P_n value for the decaying isotope i and $\lambda_i N_i(t)$ its activity.

Similarly, the total number of delayed neutrons $\bar{\nu}_d$ can be calculated as follows:

$$\bar{\nu}_d = \sum_i P_{ni} Y_i, \tag{3.7}$$

where Y_i is the yield of isotope, i.

The calculation of these expressions can be simplified in the six-group case,

$$n_{\mathrm{emit}}(t) = \bar{\nu}_d \sum_{k=1}^{6} \alpha_k \lambda_k \exp\{-\lambda_k t\}. \tag{3.8}$$

If we have constant irradiation or a long irradiation, meaning that the irradiation time is much longer than the delayed neutron emitter half-lives, then

$$n_{\mathrm{emit}}(t) = \bar{\nu}_d \sum_{k=1}^{6} \alpha_k \exp\{-\lambda_k t\}. \tag{3.9}$$

As mentioned, in practice, without the group formula, these quantities are difficult to calculate since over 100 delayed neutron precursors have been identified and P_n values are not accurately known. Moreover, accurate cumulative fission yields are quite rare except for the thermal fission of ^{235}U.

3.4.1 Importance for reactor control

In the operation of fission reactors, it turns out that β-delayed neutrons are extremely important. Although the total number of β-delayed neutrons is less than 1%, regardless of the type of reactor, they are critical in reactor operation and control. This is because if criticality relied only on prompt neutrons, control could not be maintained using external mechanisms, which would be too slow in comparison to the exponential rise in the number of neutrons. The chain reaction would either die out, or get out of control. As a result, control of the reactor power would have to be achieved by using intrinsic stability factors, such as the thermal expansion of the reactor core, or the use of materials capable of resonantly absorbing neutrons at higher temperatures. Such approaches, however, pose risks of heat damage.

Reactors are instead operated in subcritical condition, and rely on β-delayed neutrons to sustain the chain reaction while this is dying out; this still allows the use of control rods to maintain control of reactor operation. Taking into account the fact that they constitute only a small fraction of the total number of neutrons, it may be useful to consider the following example. We define the average time, Λ, between the generation of a neutron and its absorption, where $N(t)$ represents the number of neutrons at time t, and k is the multiplication factor (the absorption of one neutron generates $kN(t)$ neutrons). Then

$$N(t + \Lambda) = kN(t) \tag{3.10}$$

$$\frac{dN}{dt} = \frac{N(t + \Lambda) - N(t)}{\Lambda} = \frac{(k - 1)}{\Lambda}N(t) \tag{3.11}$$

$$N(t) = N(0)\exp\left\{\left(\frac{k - 1}{\Lambda}\right)t\right\}. \tag{3.12}$$

Equation (3.12) is also known as the point kinetics equation. Let us now consider a value for Λ of 20 μs which is typical of the prompt neutron lifetime of a thermal light water reactor (LWR). An increase of k by only 0.01% ($k = 1.0001$) would result in an increase of neutron population by a factor of 150, which is impossible to control on a time scale of 1 s (as in the accident at the Chernobyl reactor):

$$\frac{N(t)}{N(0)} = \exp\left\{\frac{1.0001 - 1}{20 \times 10^{-6}}\right\} \sim 148.4. \tag{3.13}$$

If one considers delayed neutrons, the lifetime will need to be changed to

$$\Lambda = \Lambda_{\text{fis}}(1 - B) + \Lambda_{\beta_b}B \tag{3.14}$$

$$\Lambda = 20 \times 10^{-6}(1 - 0.0068) + 12.5 \times 0.0068 = 0.085\ 02, \tag{3.15}$$

where B represents the number of beta-delayed neutrons and 12.5 is the mean beta-delayed neutron half-life in seconds.

The same calculation leads to

$$\frac{N(t)}{N(0)} = \exp\left\{\frac{1.0001 - 1}{0.085\ 02}\right\} = 1.001\ 18, \tag{3.16}$$

which makes the increase easy to control.

In other words, the delayed neutrons come a moment later, just in time to sustain the chain reaction when it is going to die out. In that regime, neutron production overall still grows exponentially, but on a time scale that is governed by the delayed neutron production, which is slow enough to be controlled by reactor control rods. The multiplication factor, k, is discussed further with regards to reactivity control in section 4.1.4.

3.4.2 Nuclear structure

From the perspective of nuclear structure, beta-delayed neutrons can provide information on the beta decay to states populated above the neutron separation energy S_n in the daughter nucleus. This information is useful for validating the results of nuclear models, used to calculate the matrix elements that connect the parent decaying state and states in the neutron-emitting daughter nucleus with the appropriate transition operator as mentioned in section 1.1.3.

States populated in beta decay are characterized by the beta strength function, defined experimentally as

$$S_{\text{exp}}(E) = \frac{P(E)}{f(Z, Q - E)T_{1/2}} = \frac{1}{ft(E)}, \tag{3.17}$$

where $P(E)$ represents the beta decay probability to the state at excitation energy E in the neutron-emitting daughter nucleus, $f(Z, Q - E)$ is the statistical rate function (or Fermi integral) that takes into account the phase space available to the leptons emitted in the process and their interaction with the Coulomb field of the daughter nucleus, $T_{1/2}$ is the half-life of the beta decay, and Q is the Q-value of the decay. The t represents the partial half-life of the decay to the level, and is calculated as

$$t = \frac{T_{1/2}}{P(E)}. \tag{3.18}$$

The ft values are conventionally given in evaluated nuclear structure databases and are quoted as $\log(ft)$ values. Please note that $P(E)$ is the feeding to the level at energy E normalized to 1.

The experimental strength $S_{\text{exp}}(E)$ can be compared with the theoretical strength defined as

$$S_{\text{theo}} = \frac{1}{6147}\left(\frac{g_A}{g_V}\right)^2 \sum_{E_f \in \Delta E} \frac{1}{\Delta E} B_{i \to f} \tag{3.19}$$

$$B_{i \to f} = \frac{1}{2J_i + 1}\left| <\psi_f \middle| O \middle| \psi_i> \right|^2, \tag{3.20}$$

where $B_{i \to f}$ represents the square of the matrix element of the beta transition operator O between the initial and final state.

The beta decay half-life $T_{1/2}$ and the P_n value are related to the beta strength according to the following relations:

$$\frac{1}{T_{1/2}} = \int_0^Q S(E)f(Z, Q - E)\mathrm{d}E \tag{3.21}$$

$$P_n = \frac{\int_{S_n}^Q S(E)f(Z, Q - E)\frac{\Gamma_n}{\Gamma_n + \Gamma_\gamma}\mathrm{d}E}{\int_0^Q S(E)f(Z, Q - E)\mathrm{d}E}. \tag{3.22}$$

These equations show the integral character of both quantities. Here Γ_n and Γ_γ are neutron and gamma widths that represent the probability of de-excitation of the states by neutron or gamma emission.

When we talk about nuclear structure in beta decay studies, conventionally, we are concerned about the comparison of the experimental data with the result of a theoretical model calculation. Here, we are concerned about the nuclear states themselves (excitation energies (E), quantum characteristics (J^{π})) and their population probabilities in the decay ($P(E)$). This can be performed by a direct comparison of $S_{\text{exp}}(E)$ with $S_{\text{theo}}(E)$ in the case when such measurements are available, for example from a combination of total absorption spectroscopy and beta-delayed neutron measurements or by comparing measurements of the integral quantities ($T_{1/2}$,P_{n}) with model predictions.

As we move further away from the stability line in the chart of nuclides, the energy window ($Q_{\beta_{\text{n}}}$) for beta-delayed neutron emission widens:

$$Q_{\beta_{\text{n}}} = Q_{\beta} - S_{\text{n}}. \tag{3.23}$$

So, the process of beta-delayed neutron emission will become more and more relevant to extract nuclear structure information and to identify and characterize the states populated in the beta decay process [41, 42].

States above the neutron separation energy S_n in the daughter nucleus can also de-excite by gamma emission. This is taken into account in the $S_{\text{theo}}(E)$ by the ratio of the gamma and neutron widths $\frac{\Gamma_{\text{n}}}{(\Gamma_{\text{n}} + \Gamma_{\gamma})}$. This is an important feature that, in the past, many theoretical calculations neglected. Most advanced theoretical calculations currently take gamma and neutron competition into account using a statistical model [43]. This can also be of great relevance in the description of multiple neutron emission processes and in general for astrophysical applications, since for example the competition reduces the number of beta-delayed neutrons predicted in theoretical calculations [44–46].

As shown by Valencia *et al* [28] and Tain *et al* [47], to evaluate the gamma–neutron competition, total absorption gamma spectroscopy measurements are needed, and the comparison of the ratio of the feedings that is followed by gamma versus the total feeding (gamma plus neutron) can be used to validate the parameters used in the statistical model. This information can also be relevant for estimating neutron capture cross-sections in very exotic nuclei since measurements are for the moment unavailable. The model estimation of neutron capture cross-sections requires the validation of the gamma widths (Γ_{γ}) deduced in the statistical model.

Another nuclear structure application is related to shape determination. In particular cases, there is a dependence of the $S_{\text{theo}}(E)$ on the shape of the parent nucleus. This can be used to infer the shape of the beta decaying state. For that, ideally, total absorption spectroscopy has been used [48–53], but some dependence can be expected in integral quantities as well [54, 55].

3.4.3 Nuclear astrophysics

It turns out that a knowledge of β-delayed neutron decays is important in some aspects of nuclear astrophysics. One of the major questions we would like to answer in this field is how and where the chemical elements are made. It was only in relatively recent times that we were able to determine the relative abundance of all the elements [56] once we had access to information on elemental abundances in stars other than the Sun. At quite an early time, after the discovery of atomic nuclei and observation of nuclear reactions, it was realized that the large amount of energy released in nuclear reactions meant that they were at the heart of element creation. It is not our purpose here to discuss the whole story. The reader is directed to [57] where the whole story, starting with H and He at the Big Bang, is laid out. Here we are concerned with the origin of the elements above iron. These nuclei are thought to be made in three processes called the slow neutron capture process (s), rapid neutron capture process (r) and the rapid proton capture process (rp). Here, we will be concerned with the r-process since beta-delayed neutron emission plays a significant role.

Both the s- and r-processes depend on neutron capture reactions but under very different conditions. In both processes, a nucleus with proton number Z and mass number A captures a neutron and is changed into a nucleus with $(Z, A + 1)$. Another neutron capture takes us to $(Z, A + 2)$ and the chain of reactions will continue until we reach an isotope which is unstable. What happens next depends on the lifetime of the unstable nucleus. In the s-process, the neutron flux is relatively low and there is a significant delay between captures. In nuclei near the line of nuclear stability, the s-process takes us towards longer-lived nuclei in general and we slowly climb in mass creating heavier nuclei as we go.

In the r-process, the neutron flux is much higher and there is a very short time between successive captures. At each step, there is competition between the beta decay and successive neutron capture. It is this competition that determines how the matter flows in the process overall. The process starts with seed nuclei around ^{56}Fe. The flux of neutrons is so high that a neutron is captured every 10^{-3} s or so. This is much shorter than the lifetimes of nuclei on or near the line of nuclear stability. Accordingly, a string of rapid neutron captures takes us away from stability. However, as we move away from stability the nuclear lifetimes decrease and beta decay competes with neutron capture. Eventually, when the neutron flux ceases, the nuclei that have been created beta decay in several steps, back to the line of stability.

As neutrons are captured in the r-process, we inevitably reach nuclei with closed neutron shells, for example, at $N = 50$, 82 and 126. These nuclei with low neutron capture cross-sections are called waiting point nuclei because they have a higher probability of beta decaying towards stability rather than capturing another neutron. In the s-process many stable nuclei with closed neutron shells are created. In contrast the r-process creates nuclei with magic numbers with masses some 10 amu (atomic mass units) lower than these s-process peaks. Accordingly, in the typical observed abundance plots, we see peaks at the magic numbers from the s-process and in each case a peak at lower mass following the decay of the waiting point nuclei in the r-process. As indicated earlier as the r-process proceeds, we reach nuclei that can decay with the emission of beta-delayed neutrons. Once we reach this stage, the beta decay process, since it involves beta-delayed neutron emission, leads to final nuclei of slightly lower mass changing the final abundance.

Where does the r-process occur and can we calculate the resulting abundances? In answer to the first question, it must involve a process or processes where there is an enormous flux of neutrons. Indeed, it demands a flux of neutrons greater than 3×10^{20} cm^{-3} [57]. Where in the Cosmos might this happen? Two favoured possibilities are core collapse supernovae and neutron star mergers. However, in both cases it is difficult to pin down the details [58]. For example, do neutron star mergers occur early enough in the history of the Universe to explain the abundances observed in the oldest stars?

Turning to the second question we need, as input to the calculations, all of the (n,γ) cross-sections and the decay properties of the heavy neutron-rich nuclei involved. In particular we require the half-lives and P_n values for beta-delayed neutron emission since it modifies the beta decay path by lowering the A values. The P_{1n}, P_{2n} values are also important because they contribute neutrons to overall neutron flux. Because our current knowledge of beta-delayed neutron decays is limited, our calculations of abundances must rely on theory, for example, the Hauser–Feshbach statistical model to provide cross-sections [43]. This is based on the use of average values for nuclear properties. It is clear that we need much more information about neutron-rich nuclei that decay by beta-delayed neutron emission, if we are to do better.

3.4.4 Measurement methods

As discussed earlier, beta-delayed neutron emission probability is defined as

$$P_{1n} = \frac{N_{1n}}{N_{decays}}, \tag{3.24}$$

where N_{decays} represents the number of beta decays and N_{1n} stands for the number of beta decays that are followed by one neutron emission. Similar expressions can be defined for the multiple neutron emission processes as well, i.e. P_{xn}, where x represents the neutron multiplicity.

There are different methods to determine these quantities experimentally. In the following, we will discuss briefly the most relevant techniques and mention the most common challenges. More details can be found in a recent review by Dimitriou *et al* [40] that updates the related work by Rudstam *et al* [59].

Most of the measurements found in the literature are based on methods that count the number of neutrons using moderated neutron detectors. In those set-ups, a medium composed of a highly hydrogenated material such as paraffin wax or high-density polyethylene is used as a moderator. Note that the high hydrogen content is used to maximize the number of elastic interactions with neutrons, that reduce the neutrons' kinetic energies and finally thermalize them.

The first such detectors were built using BF_3 filled proportional tubes, but currently the use of detectors filled with ^3He is more common. Both ^3He and ^{10}B have a large neutron capture cross section. The nuclear reaction used in the ^3He detectors is the following:

$$n + {}^3He \rightarrow {}^3H + {}^1H + 765\ keV \tag{3.25}$$

where the 765 keV released is shared among the charged particles created.

The advantage of ^{3}He-based detectors is the large detector efficiency, the almost non-existent low energy threshold and the small gamma detection sensitivity. Examples of such systems are NERO (now HABANERO) [60] at NSCL (USA), LOENIE at Lohengrin Institut Laue Langevin (France) [61], 3Hen [62] at HFIBF Oak Ridge (USA), BELEN [63] developed for FAIR (Germany) and used at the University of Jyväskylä (Finland) and GSI (Germany), TETRA [64] at ALTO (France), etc. The largest such system is BRIKEN [65, 66] at RIKEN constructed in the framework of a large international collaboration combining detectors from 3Hen, BELEN, and RIKEN.

The major concern with ^{3}He-based detector arrays is the dependence of the efficiency of the array on the neutron energy. The effect depends on the geometry of the set-up, i.e. the amount of moderator material and the spatial distribution of the counters. Modern designs try to maximize the efficiency and, at the same time, try to reduce the energy dependence. This can be studied with Monte Carlo simulations [65, 66].

We now turn to briefly describing the most common measurement methods for determining P_n. Nowadays, one of the most common methods is the 'β-neutron' coincidence method (see [40, 67, 68] for more details). As the name indicates, it is based on counting the number of β-neutron coincidences and comparing it with the number of beta particles:

$$P_{1n} = \frac{1}{\varepsilon_{1n}} \frac{N_{\beta_{1n}}}{N_\beta}, \tag{3.26}$$

where ε_{1n} is the one neutron detection efficiency, $N_{\beta_{1n}}$ is the number of time-correlated β-1n events and N_β is the number of beta decays. The main assumption in the formula is that the beta efficiency of the beta detector used in the coincidence measurements is constant for all energies and cancels out. This is not always a correct assumption [67, 68]. So, more precisely the equation should be

$$P_{1n} = \frac{\langle \varepsilon_\beta \rangle}{\langle \varepsilon^*_\beta \varepsilon_n \rangle} \frac{N_{\beta_{1n}}}{N_\beta}, \tag{3.27}$$

where $\langle \varepsilon_\beta \rangle$ is the average beta efficiency over the full energy range and $\langle \varepsilon^*_\beta \varepsilon_n \rangle$ is the product of the average beta efficiency for states populated in the beta decay above the neutron separation energy and the average neutron detection efficiency. If $\langle \varepsilon_\beta \rangle = \langle \varepsilon^*_\beta \rangle$ then we obtain the earlier expression, equation (3.26).

Another common method is the separated simultaneous neutron and beta counting (the 'nβ' method):

$$P_{1n} = \frac{\varepsilon_\beta}{\varepsilon_{1n}} \frac{N_{1n}}{N_\beta}. \tag{3.28}$$

This method, like the earlier one, requires a careful determination of the relevant efficiencies, and the study of possible systematic uncertainties related to the effect of the different beta decay Q_β values on the efficiencies both for the cases to be measured and for the calibration sources.

Another common measurement method is the '$\gamma\gamma$' method, which relies on the detection of gamma-rays following the beta decays of the relevant nuclei (so, detecting gammas from the precursor, to estimate the total number of precursor nuclei, and gamma-rays from the beta decay of the beta-delayed neutron daughter to obtain the number of beta-delayed neutrons). The method assumes that we know the absolute beta decay branches for these decays or, in other words, that we know the absolute number of gammas emitted per 100 beta decays of the decays of interest. Then we can write

$$\frac{P_{1n}}{1 - P_{1n}} = \frac{\varepsilon_\gamma^A I_\gamma^{\text{abs},\, A}}{\varepsilon_\gamma^{A-1} I_\gamma^{\text{abs},\, A-1}} \frac{N_\gamma^{A-1}}{N_\gamma^A}, \tag{3.29}$$

where ε_γ^A and $\varepsilon_\gamma^{(A-1)}$ are the gamma efficiencies for the relevant gamma-rays, $I_\gamma^{(\text{abs},\, A)}$ and $I_\gamma^{(\text{abs},\, A-1)}$ are the absolute gamma emission branches of the respective beta decays, and $N_\gamma^{(A-1)}$ and N_γ^A are the number of counts of the respective gamma-rays.

There is a family of more recent methods that is also worth mentioning. These methods are based on ion counting, as for example detecting the recoils after neutron emission in a Paul trap [69], observing the recoil tracks in an optical time-projection chamber [70] or just counting ions in a heavy ion storage ring [71] or multi-reflection time of flight spectrometer [72]. The majority of these methods have been born as a result of recent technical developments related to radioactive ion beam facilities.

In some applications the neutron energy spectrum is also needed. The most common measurement techniques are the following: ^{3}He spectrometers, gaseous proton recoil spectrometers and time of flight spectrometers. In the following, we only present very briefly the most relevant characteristics of each technique (the reader will find more detail and examples in the recent review by Dimitriou *et al* [40] and the references therein):

- **^{3}He spectrometers:** The method is based on the detection of fast neutrons using the earlier mentioned $n + {}^3\text{He} \rightarrow {}^3\text{H} + p$ reaction, that will produce an energy peak shifted with respect to the thermal neutron peak position by an energy proportional to the detected neutron energy. The limit of the measurable energy in these detectors is about 3 MeV.
- **Proton recoil spectrometers:** They are based on the detection of protons elastically scattered by neutrons. These detectors use methane or H_2 gas at high pressure and act as proportional counters.
- **Time of flight (TOF) spectrometers:** these systems are based on the detection of neutrons in coincidence with beta particles. One measures a time interval between a start signal given by the beta particle detector and a stop signal provided by a neutron detector, that can be constructed from a plastic scintillator with high H content or a liquid scintillator. Since the distance between the implantation point where the beta decay occurs and the neutron

detectors is well defined, the velocity and thus the kinetic energy of the neutrons can be deduced from the time measurement. The distance and the timing resolution of the set-up determines the energy resolution. Examples of such detectors are TONNERRE [73] used at CERN and GANIL, TOF [74] developed at NSCL and, more recently, DESCANT [75, 76], VANDLE [77] and MONSTER [78].

References

[1] Chadwick J 1932 The existence of a neutron *Proc. R. Soc.* **136** 692

[2] Curie M and Joliot F 1934 Un nouveau type de radioactivité *C. R. Hebd. Seances Acad. Sci.* **198** 254–6

[3] Joliot F and Curie I 1934 Artificial production of a new kind of radio-element *Nature* **133** 201–2

[4] Guerra F, Leone M and Robotti N 2012 The discovery of artificial radioactivity *Phys. Perspect.* **14** 33–58

[5] Hahn O and Strassman F 1939 Concerning the existence of alkaline earth metals resulting from neutron irradiation of uranium *Naturwissenschaften* **27** 11–5

[6] Meitner L and Frisch O R 1939 Disintegration of uranium by neutrons: a new type of nuclear reaction *Nature* **143** 239–40

[7] Gamow G 1930 Mass defect curve and nuclear constitution *Proc. R. Soc.* **126** 632–44

[8] Way K and Wigner E P 1948 The rate of decay of fission products *Phys. Rev.* **73** 1318

[9] Hardy J C, Jonson B and Hansen P G 1978 The essential decay of pandemonium: β-delayed neutrons *Nucl. Phys.* A **305** 15–28

[10] Akkoyun S *et al* 2012 AGATA—Advanced GAmma Tracking Array *Nucl. Instrum. Meth.* A **668** 26–58

[11] Lee I Y, Deleplanque M A and Vetter K 2003 Developments in large gamma-ray detector arrays *Rep. Prog. Phys.* **66** 1095

[12] Algora A *et al* 2003 Fine structure of the Gamow–Teller resonance revealed in the decay of ^{150}Ho 2^- isomer *Phys. Rev.* C **68** 034301

[13] Karny M *et al* 1997 Coupling a total absorption spectrometer to the GSI on-line mass separator *Nucl. Instrum. Meth.* B **126** 411–5

[14] Rubio B, Gelletly W, Algora A, Nacher E and Tain J L 2017 Beta decay studies with total absorption spectroscopy and the Lucrecia spectrometer at ISOLDE *J. Phys.* G: Nucl. Part. Phys. **44** 084004

[15] Algora A, Tain J L, Rubio B, Fallot M and Gelletly W 2021 Beta-decay studies for applied and basic nuclear physics *Eur. Phys. J.* A **57** 85

[16] Tain J L and Cano-Ott D 2007 Algorithms for the analysis of β-decay total absorption spectra *Nucl. Instrum. Meth.* A **571** 728–38

[17] Tain J L and Cano-Ott D 2007 The influence of the unknown de-excitation pattern in the analysis of β-decay total absorption spectra *Nucl. Instrum. Meth.* A **571** 719–27

[18] Yoshida T, Tachibana T, Storrer F, Oyamatsu Y and Katakura J 1999 Possible origin of the gamma-ray discrepancy in the summation calculations of fission product decay heat *J. Nucl. Sci. Technol.* **36** 135–42

[19] Yoshida T *et al* 2007 Assessment of fission product decay data for decay heat calculations, *Technical Report* NEA/WPEC 25 1 NEA

[20] Nichols A L 2006 Summary report of the second consultants' meeting on beta-decay and decay heat, *Technical Report* INDC(NDS) 0499 IAEA

[21] Gupta M, Kellett M A, Nichols A L and Bersillon O 2010 Decay heat calculations: Assessment of fission product decay data requirements for Th/U fuel, *Technical Report* INDC(NDS)-0577 IAEA

[22] Äystö J 2001 Development and applications of the IGISOL technique *Nucl. Phys.* A **693** 477–94

[23] Moore I D *et al* 2013 Towards commissioning the new IGISOL-4 facilitys *Nucl. Instrum. Meth.* B **317** 208–13

[24] Dendooven P 1997 The development and status of the IGISOL technique *Nucl. Instrum. Meth.* B **126** 182–9

[25] Kolhinen V S *et al* 2004 JYFLTRAP: a cylindrical Penning trap for isobaric beam purification at IGISOL *Nucl. Instrum. Meth.* A **528** 776–87

[26] Algora A *et al* 2010 Reactor decay heat in ^{239}Pu: Solving the γ discrepancy in the 4–3000-s cooling period *Phys. Rev. Lett.* **105** 202501

[27] Jordan D *et al* 2013 Total absorptions study of the β decay of $^{102,\,104,\,105}$Tc *Phys. Rev.* C **87** 044318

[28] Valencia E *et al* 2017 Total absorption γ-ray spectrscopy of the β-delayed neutron emitters ^{87}Br, ^{88}Br and ^{94}Rb *Phys. Rev.* C **95** 024320

[29] Rice S *et al* 2017 Total absorption spectroscopy study of the β decay of ^{86}Br and ^{91}Rb *Phys. Rev.* C **96** 014320

[30] Zakari-Issoufou A A *et al* 2015 Total absorption spectroscopy study of ^{92}Rb decay: a major contributor to reactor antineutrino spectrum shape *Phys. Rev. Lett.* **115** 102503

[31] Tobias A 1989 Derivation of decay heat benchmarks for 235 U and 239 Pu by a least-squares fit to measured data, *Technical Report* RD/B/6210/R89 CEGB

[32] Dickens J K, Love T A, McConnell J W and Peele R W 1980 Fission-product energy release for times following thermal-neutron fission of ^{235}U between 2 and 14000 *Nucl. Sci. Eng.* **74** 106–29

[33] Dickens J K, Love T A, McConnell J W and Peelle R W 1981 Fission-product energy release for times following thermal-neutron fission of plutonium-239 and plutonium-241 between 2 and 14000 *Nucl. Sci. Eng.* **78** 126–46

[34] Roberts R B, Meyer R C and Wang P 1939 Further observations on the splitting of uranium and thorium *Phys. Rev.* **55** 510

[35] Booth E T, Dunning J R and Slack F G 1939 Delayed neutron emission from uranium *Phys. Rev.* **55** 876

[36] Bohr N and Wheeler J A 1939 The mechanism of nuclear fission *Phys. Rev.* **56** 426

[37] Roberts R B, Hafstad L R, Meyer R C and Wang P 1939 The delayed neutron emissions which accompanies fission of uranium and thorium *Phys. Rev.* **55** 664

[38] Feather N 1939 The time involved in the process of nuclear fission *Nature* **143** 597–8

[39] Mccutchan E 2018 Nuclear data sheets for A=136 *Nucl. Data Sheets* **152** 331–667

[40] Dimitriou P *et al* 2021 Development of a reference database for beta-delayed neutron emission *Nucl. Data Sheets* **173** 144–238

[41] Xu Z Y *et al* 2023 A Rosetta stone for decays of r-process nuclei *Phys. Rev. Lett.* **131** 022501

[42] Cox I *et al* 2024 Proton shell gaps in N=28 nuclei from the first complete spectroscopy study with FRIB decay station initiator *Phys. Rev. Lett.* **132** 152503

[43] Hauser W and Feschbach H 1952 The inelastic scattering of neutrons *Phys. Rev.* **87** 366

[44] Yokoyama R *et al* 2019 Strong one-neutron emission from two-neutron unbound states in β decays of the r-process nuclei $^{86,\,87}$Ga *Phys. Rev.* C **100** 031302
[45] Kawano T *et al* 2008 Calculation of delayed-neutron energy spectra in a quasiparticle random-phase approximation Hauser–Feschbach model *Phys. Rev.* C **78** 054601
[46] Möller P, Mumpower M R, Kawano T and Myers W D 2019 Nuclear properties for astrophysical and radioactive-ion-beam applications (II) *At. Data Nucl. Data Tables* **125** 1–192
[47] Tain J L *et al* 2015 Enhanced γ-ray emission from neutron unbound states populated in β decay *Phys. Rev. Lett.* **115** 062502
[48] Poirier E *et al* 2004 B(GT) strength from β-decay measurements and inferred shape mixing in ^{74}Kr *Phys. Rev.* C **69** 034307
[49] Nácher E *et al* 2004 Deformation of the $N{=}Z$ nucleus ^{76}Sr using β-decay studies *Phys. Rev. Lett.* **92** 232501
[50] Pérez-Cerdán A B 2013 Deformation of Sr and Rb isotopes close the $n = z$ line via β-decay studies using the total absorption technique *Phys. Rev.* C **88** 014324
[51] Briz J A *et al* 2015 Shape study of the $N{=}Z$ nucleus ^{72}Kr via β decay *Phys. Rev.* C **92** 054326
[52] Estévez Aguado M E *et al* 2015 Shapes of $^{192,\,190}$Pb ground states from β-decay studies using the total-absorption technique *Phys. Rev.* C **92** 044321
[53] Algora A *et al* 2021 Total absorption gamma-ray spectroscopy study of the β-decay of ^{186}Hg *Phys. Lett.* B **819** 136438
[54] Sarriguren P, Algora A and Pereira J 2014 Gamow–Teller response in deformed even and odd neutron-rich Zr and Mo isotopes *Phys. Rev.* C **89** 034311
[55] Sarriguren P and Pereira J 2010 β-decay properties of neutron-rich Zr and Mo isotopes *Phys. Rev.* C **81** 064314
[56] Suess H E and Urey H C 1956 Abundances of the elements *Rev. Mod. Phys.* **28** 53
[57] Rolfs C and Rodney W S 1988 *Cauldrons in the Cosmos* (University of Chicago Press)
[58] Mumpower M, Surman R, McLaughlin G C and Aprahamian A 2016 The impact of individual nuclear properties on r-process nucleosynthesis *Prog. Part. Nucl. Phys.* **86** 86–126
[59] Rudstam G, Aleklett K and Sihver L 1993 Delayed-neutron branching ratios of precursors in the fission product region *At. Data Nucl. Data Tables* **53** 1–22
[60] Pereira J *et al* 2010 The neutron long counter NERO for studies of β-delayed neutron emission in the r-process *Nucl. Instrum. Meth.* A **618** 275–83
[61] Mathieu L *et al* 2012 New neutron long-counter for delayed neutron investigations with the LOHENGRIN fission fragment separator *J. Instrum.* **7** P08029
[62] Grzywacz R *et al* 2014 Hybrid-3Hen—new detector gammas and neutrons *Acta Phys. Pol.* B **45** 217
[63] Agramunt J *et al* 2016 Characterization of a neutron-beta counting system with beta-delayed neutron emtitters *Nucl. Instrum. Meth.* A **807** 69–78
[64] Testov D *et al* 2016 The ^{3}He long-counter TETRA at the ALTO ISOL facility *Nucl. Instrum. Meth.* A **815** 96–103
[65] Tarifeno-Saldivia A *et al* 2017 Conceptual design of a hybrid neutron-gamma detector for study of β-delayed neutrons at the RIB facility of RIKEN *J. Instrum.* **12** P04006
[66] Tolosa-Delgado A *et al* 2019 Commissioning of the BRIKEN detector for the measurement of very exotic β-delayed neutron emitters *Nucl. Instrum. Meth.* A **925** 133–47
[67] Agramunt J *et al* 2014 New beta-delayed neutron measurements in the light-mass fission group *Nucl. Data Sheets* **120** 74–7

[68] Caballero-Folch R *et al* 2018 First determination of β-delayed multiple neutron emission beyond A =100 through direct neutron measurements: The P_{2n} value of ^{136}Sb *Phys. Rev.* C **98** 034310

[69] Yee R M *et al* 2013 β-delayed neutron spectroscopy using trapped radioactive ions *Phys. Rev. Lett.* **110** 092501

[70] Mianowski S *et al* 2010 Imaging the decay of ^{8}He *Acta Phys. Pol.* B **41** 449

[71] Evdokimov A *et al* 2012 An alternative approach to measure β-delayed neutron emission *XII Int. Symp on Nuclei in the Cosmos*

[72] Miskun I *et al* 2019 A novel method for the measurements of half-lives and decay branching ratios of exotic nuclei *Eur. Phys. J.* A **55** 148

[73] Buta A *et al* 2000 TONNERRE: an array for delayed-neutron decay spectroscopy *Nucl. Instrum. Meth.* A **455** 412

[74] Harkewicz R, Morrissey D J, Brown B A, Nolen J A, Orr N A, Sherrill B M, Winfield J S and Winger J A 1991 β-decay branching ratios of the neutron-rich nucleus ^{15}B *Phys. Rev.* C **44** 2365

[75] Bildstein V *et al* 2013 Comparison of deuterated and normal liquid scintillators for fast-neutron detection *Nucl. Instrum. Meth.* A **729** 188–97

[76] Garrett P E *et al* 2013 DESCANT—the deuterated scintillator array for neutron tagging *Hyperfine Interact.* **225** 137–41

[77] Peters W A *et al* 2016 Performance of the Versatile Array of Neutron Detectors at Low Energy (VANDLE) *Nucl. Instrum. Meth.* A **836** 122–33

[78] Martínez T *et al* 2014 MONSTER: a TOF spectrometer for β-delayed neutron spectroscopy *Nucl. Data Sheets* **120** 78–80

IOP Publishing

Nuclear Data
Applications to society and industry
David G Jenkins

Chapter 4

Fission reactor monitoring

Andrew Petts

This chapter will explore the practical applications of nuclear data within the nuclear industry on an operational reactor. Topics covered will be routine, regular reactor core assessment, fuel performance and monitoring for nuclear security and safeguards.

Concepts: Reactor monitoring, nuclear security, nuclear safeguards, nuclear fuel performance

Within the nuclear industry, the monitoring and assessment of the reactor core is crucial. A regular and rigorous monitoring and assessment regime helps ensure that the reactor operates within its design specifications and remains well within its safety envelope. This ensures that the risk of a radiological release that could cause significant harm to the public is minimized and kept *as low as reasonably practicable* (ALARP). The ALARP principle to which nuclear installations operate results in this risk being managed to a 1 in 10 000 year return frequency. Routine monitoring of the fissile material inventory is also required to ensure compliance with International Atomic Energy Agency (IAEA) safeguard systems for non-proliferation and to help maintain nuclear security. This chapter will first discuss how nuclear data play a key part in routine reactor monitoring and compliance calculations. It will then explore how nuclear data are used in the provision of nuclear safeguards before discussing how data from reactor monitoring can be used to enhance nuclear security by means of providing information in support of detecting unsanctioned nuclear tests.

4.1 Routine reactor assessments

For all credible faults and hazards (those which have been determined to have a return frequency greater than 1 in 1 000 000 years), the safety case for the operating facility must demonstrate how the reactor can be safely shutdown and cooled to reduce the risk of a radiological release to below the ALARP threshold. How this is achieved is referred to as a *line of protection* (LOP). For 'frequent' faults (those with a return frequency greater than 1 in 1000 years), two LOPs are required to be

doi:10.1088/978-0-7503-5102-7ch4
© IOP Publishing Ltd 2025. All rights, including for text and data mining (TDM), artificial intelligence (AI) training, and similar technologies, are reserved.

demonstrated. The safety case will define the plant and operating conditions required to secure the LOPs. In practice, a reactor facility will ensure that the LOPs are available by complying with *technical specifications* (Tech Specs) and *limiting conditions of operation* (LCOs), which are distilled from the requirements of the safety case and define acceptable envelopes of safe operation. Most Tech Specs and LCOs have measurable parameters that fall under a periodic surveillance to demonstrate that the facility is operating within the bounds of the safety case. However, some parameters important to the safety case cannot be measured directly. These parameters are mostly associated with the nuclear fuel itself, such as fuel pellet temperature and fuel pin internal pressure. In order to ensure that these parameters are within the assumptions of the safety case, routine reactor physics calculations are performed to calculate their values and track their evolution. These routine assessments also allow for the current reactivity balance and pin inventories to be calculated, which are used for shutdown margin and decay heat assessments.

4.1.1 Core follows and regular assessment routes

The calculations within the regular assessment route (RAR), or core follow, are performed periodically and are point-to-point: the previous core state is updated with the latest measured operational plant data to produce a revised, up-to-date core state. This allows the reactor physicist to 'follow' the evolution of the reactor core. The revised core state can then be used to model and assess the effect of potential transients or faults for the current operating regime.

The RAR takes in measured plant data such as reactor temperatures, coolant flows and control rod positions to update the previous core state. The primary purpose of the route is to solve (an approximation to) the neutron transport equation and use the solutions of this equation to calculate fuel irradiations, fuel pin temperatures and core reactivity. The neutron transport equation can be written in terms of neutron flux as

$$\begin{aligned}&\frac{1}{\nu}\frac{\partial}{\partial t}\phi(\vec{r}, E, \hat{\Omega}, t) + \hat{\Omega}\cdot\nabla\phi(\vec{r}, E, \hat{\Omega}, t) + \Sigma_t(\vec{r}, E)\phi(\vec{r}, E, \hat{\Omega}, t) = \\ &\int_{4\pi} \mathrm{d}\hat{\Omega}'\int_0^{\infty} \mathrm{d}E'\Sigma_s(E' \to E, \hat{\Omega}' \to \hat{\Omega})\phi(\vec{r}, E, \hat{\Omega}', t) + s(\vec{r}, E, \hat{\Omega}, t).\end{aligned} \tag{4.1}$$

This complicated equation is effectively a neutron balance within a unit volume. To the right of the equation are terms associated with in-scatter of neutrons and a source term describing neutron production. The terms on the left-hand side of the equation deal with absorption, out-scatter and net leakage. The neutron transport equation is extremely difficult to solve due to it having both integral and differential components and several independent variables: position (r), energy (E), direction (Ω, ϕ) and time (t). For this reason, the equation must be simplified. The first approximation recognizes that in most reactor physics calculations, angular dependence of the neutron flux is not required. By integrating equation (4.1) over all angles, one obtains the equation of continuity:

$$\frac{1}{\nu}\frac{\partial}{\partial t}\phi(\vec{r}, E, t) + \nabla \cdot \vec{J}(\vec{r}, E, t) + \Sigma_t(\vec{r}, E)\phi(\vec{r}, E, t) = \int_0^{\infty} \Sigma_s(E' \to E)\phi(\vec{r}, E', t)\mathrm{d}E' + S(\vec{r}, E, t). \tag{4.2}$$

This equation has two unknown components, ϕ and $\boldsymbol{J}$ the neutron current, and so cannot be solved. Further simplification is therefore required. By using Fick's law, which is used in gas kinetics, the neutron current is assumed to be proportional to the gradient of the flux. Equation (4.2) becomes the energy-dependent diffusion equation:

$$\frac{1}{\nu}\frac{\partial}{\partial t}\phi(\vec{r}, E, t) - D(\vec{r}, E)\nabla^2\phi(\vec{r}, E, t) + \Sigma_t(\vec{r}, E, t)\phi(\vec{r}, E, t) = \int_0^{\infty} \mathrm{d}E'\Sigma_s(E' \to E)\phi(\vec{r}, E, t) + S(\vec{r}, E, t) \tag{4.3}$$

with $\boldsymbol{D}$ being the material-dependent diffusion coefficient. The final simplification can be obtained by removing the energy dependence. Integrating over all energies, equation (4.3) becomes the one-group diffusion equation:

$$\frac{1}{\nu}\frac{\mathrm{d}\phi}{\mathrm{d}t} = \nu\Sigma_f\phi - \Sigma_a\phi + D\nabla^2\phi. \tag{4.4}$$

The absorption and production cross sections, Σ_a and Σ_f, have material and energy dependencies. A whole-core solution to the physics problem would be too large to calculate routinely. For a standard pressurized water reactor (PWR) or advanced gas-cooled reactor (AGR) problem, to solve using a whole-core model would require around 100 terabytes of data per calculation. Consider that each fuel pin contains fuel, clad and moderator, and that there are around 100 000 fuel pins per reactor and around 100 different nuclides to consider. Cross-section data are energy dependent and their detailed behaviour is described with around 10 000 energy groups. Due to the large number of variables, simplification is required to make the calculation more practicable.

The general approach used by most reactor physics practitioners throughout the world is to use a combination of lattice and whole-core reactor codes. Lattice codes provide deterministic solutions to the neutron transport equation. Examples of lattice codes currently in use are the UK-developed WIMS [1] code and CASMO-5 [2] developed by Studsvik. The physics problem is confined to a 'cell', rather than the whole reactor. The cell size is calculation-dependent, but usually a fuel assembly or fuel pin. The energy and material dependencies are first treated separately before being combined to obtain cross-section data for the cell/fuel assembly. A detailed energy flux solution is first obtained for a simplified geometry. For example, the geometry of the fuel assembly may be simplified from a 17×17 fuel pin array down to a geometry for a single pin consisting of fuel, moderator and clad. Calculations are performed using multi-group libraries of evaluated nuclear data such as ENDF/B-VII [3], JEFF-2.2 [4] or JENDL-3.2 [5]. Energy-dependent cross sections from these calculations can then be grouped, to say six energy groups from the 172 or 281 that may have been initially used in the lattice calculation. A more detailed geometric calculation is then performed using the grouped energy cross sections. The final step for the lattice code calculation is to homogenize each pin cell to provide smeared cross sections representing the average properties of the assembly. The fuel assembly has now been reduced to a homogenized paste with no spatial data. These averaged cross sections are produced for all fuel types and assemblies within the reactor and are often produced for a range of burn-ups, temperatures and moderator conditions. Whole-core reactor physics codes are then used to solve the diffusion approximation to the neutron transport equation. UK operators use the PANTHER [6] reactor physics code for these calculations, but CASMO-simulate is widely used and often the whole core code is vendor-specific. Cross sections from the lattice code calculation are mapped to each reactor location and used to solve for the flux distribution. The flux solutions are then used in conjunction with the homogenized cross sections to calculate data such as assembly fission rates, pin ratings and fuel temperatures.

4.1.2 Pin inventory and pin pressure assessments

The purpose of nuclear safety is to ensure that the risk of radiological release is minimized and is at all times ALARP. The primary barrier to fission product release is the fuel pin itself. Fuel pins usually consist of a stainless steel alloy and contain the nuclear fuel pellets. During operation the fissile material will produce fission products. The fission process balances mass and energy. For ^{235}U fission, the process is described by

$$^{235}\mathrm{U} + \mathrm{n} \rightarrow {}^{x}\mathrm{A} + {}^{234-x}\mathrm{B} + 2\mathrm{n}. \tag{4.5}$$

Within the fission process, on average around two neutrons are produced which sustain the chain reaction. The other main products will have a total mass of 234 atomic units, conserving the total mass. The mass of the fission products vary, but are distributed bimodally with one peak around 90–100 atomic mass units and the other around 130–140 atomic mass units. Figure 4.1 shows independent fission yields for neutron-induced fission of ^{235}U where this bimodal distribution can be observed. Due to the high temperatures at which the reactor core operates, the fission products are often gaseous. An obvious fault that could occur is that the internal pressure inside the fuel pin becomes large enough to rupture the fuel cladding and allow the release of fission products into the primary coolant. In order to maintain primary barrier integrity there is an operation limit on internal pin pressure. Because this cannot be directly measured during normal operation, it is inferred by calculation. This calculation uses as inputs the local neutron flux and cross sections from the lattice and whole-core calculations described in section 4.1.1. Inventories can be calculated using modules within codes such as PANTHER or using stand-alone codes such as FISPIN [7]. This information is combined with measured parameters such as reactor temperatures and nuclear data parameters such as fission

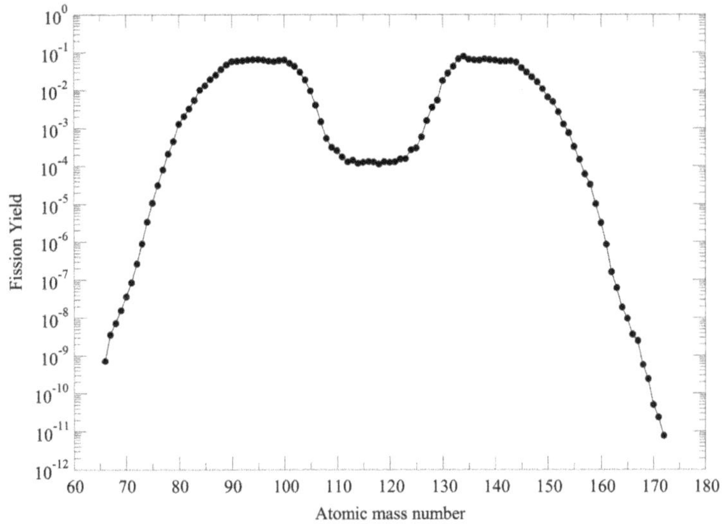

Figure 4.1. Fission yield plot for ^{235}U neutron-induced fissions using JENDL-5 [5] nuclear cross-section data at typical thermal neutron energies. This plot illustrates the bimodal nature of the mass distribution. The exact profile will differ depending on neutron energy.

yields and decay constants. The calculation outputs pin inventories (fission product inventories) and uses modelled fuel temperatures to infer the internal pressure. The pin inventories are dependent upon the time at power—the longer a reactor operates the more fission products are produced. In general, this results in pin pressures increasing over the course of a reactor run. The user then checks that the calculated pin pressures are below the allowable LCO limits. If margin to the LCO limits are too close then advice is given to manage the internal pin pressures.

4.1.3 Decay heat calculations

Decay heat is perhaps one of the most important considerations for operational nuclear reactors (see more on the underlying physics of decay heat in section 3.3). It is often cited as the reason why nuclear technology is considered 'special'—even when shutdown, there is a heat source present that must be dealt with. This heat source is present because the fission products produced during reactor operation will themselves be radioactive. The majority of the fission products will undergo β^- decay (see section 1.1.3). During these decay processes, heat is produced as radiation interacts with matter. During routine core assessments, fission product inventories are calculated, as discussed in section 4.1.2.

As well as being used to assess internal pin pressure, the pin inventories are used to determine the decay heat as a function of time. The decay heat levels vary over time due to a number of factors. The initial fission product inventories depend upon time at power—in general, the longer a reactor runs, the greater the initial inventory. There is also an energy dependence—the power at which a reactor is operating

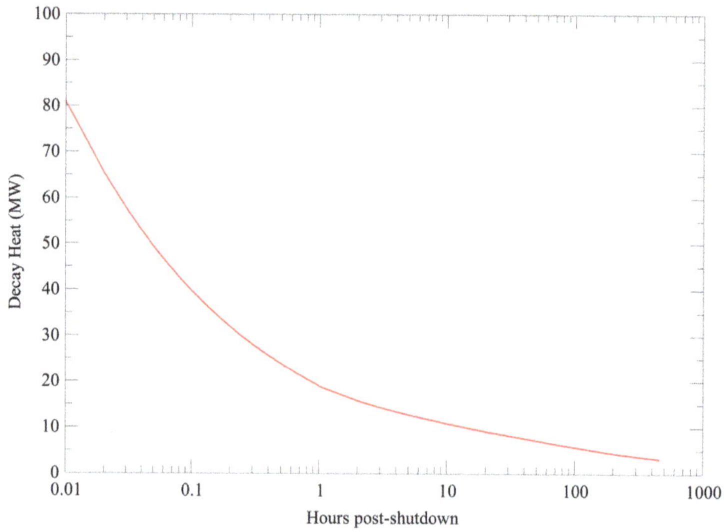

Figure 4.2. Typical reactor decay heat of an AGR following a nominal 4 month, full power run at 1550 MW. Decay heat is approximately 7% of reactor power at trip and decays exponentially with time. Even after 20 days shutdown, it can be seen that there is 3 MW of heat production which must be removed.

subtly affects production cross sections and so individual fission product inventories change depending upon the operating history of the reactor. Finally, the make-up of the pin inventory changes over time due to the varying half-lives of the fission products. The fact that decay heat changes over time is an important aspect as the decay heat level determines the minimum cooling requirements. Figure 4.2 shows a typical decay heat curve for an AGR which has operated at full load for a nominal run. It can be seen that immediately post-shutdown there is around 80 MW of heat still being produced, and even after 24 h around 10 MW is still available. Although this figure is specifically derived from AGR data, the curve is representative of any large-scale reactor that uses low-enrichment uranium as its fuel source. It is important that there is enough cooling plant in service to ensure that the reactor heat is safely removed. Should this heat not be removed it could ultimately lead to increased fuel temperatures which may either damage or melt the fuel, leading to a radiological release. During planned shutdowns, the cooling plant requires routine maintenance. As decay heat reduces with time, so does the minimum cooling requirement. The times at which plant can be released for maintenance are set on decay heat levels. This ensures that sufficient cooling capability remains as plant is taken out of service. Accuracy of these data going into the calculations is key as the uncertainties must be factored into the assessment. Large uncertainties can result in conservative and pessimistic values calculated which ultimately delays plant release and increases overall outage duration. However, a degree of pessimism is always applied and bounding levels used so as plant is not released before there is a high level of confidence that the decay heat is low enough to avoid any significant risk of fuel damage.

4.1.4 Reactivity management, shutdown margin and xenon override

The effective neutron multiplication factor, k_{eff} is a measure of the number of neutrons in the current generation versus those in the previous generation. It can be thought of in terms of neutron production mechanisms versus mechanisms associated with absorption and losses. This model leads to the *six-factor formula* for k_{eff},

$$k_{\text{eff}} = \varepsilon p f \eta P_{\text{T L}} P_{\text{FL}}. \tag{4.6}$$

Here, ε represents the *fast fission factor*. Fast fission of ^{238}U will increase the number of fast neutrons in the reactor. ε represents the increase above those occurring from the thermal fission of ^{235}U. Because the reactor is finite in size, there will be some leakage of fast neutrons from the active core. P_{FL} accounts for the leakage of fast neutrons and represents the surviving fraction, i.e those neutrons which *do not* leak from the core. This factor is maximized by optimizing the geometry of the reactor and improving reflection at the active core edges. Neutrons which have not leaked from the edge of the core may be lost to resonance capture in ^{238}U. P is referred to as the *resonance escape probability* and represents the fraction of surviving neutrons which are not captured by ^{238}U and undergo thermalization. After thermalization, some neutrons will be lost to leakage. $P_{\text{T L}}$ represents the fraction of neutrons which *do not* leak from the core before undergoing absorption. The *thermal utilization factor*, f accounts for the fraction of surviving neutrons absorbed by the fissile ^{235}U. Finally, the *thermal reproduction factor*, η represents the gain in neutrons that result from thermal fission.

As part of the routine assessment discussed in section 4.1.1, core reactivity is calculated. This is important because it is a fundamental requirement to be able to shutdown a nuclear reactor and for it to remain sub-critical should a fault occur. The level of reactivity clearly impacts how easy this is to achieve. Shutdown and hold-down is managed by the insertion of the primary shutdown system (PSD) which normally consists of boronated stainless steel rods which act as strong neutron absorbers. The PSD is designed with significant redundancy to account for partial failure (in the unlikely event that some rods do not insert) or for routine maintenance of specific control rods which would render them unavailable to insert during the fault.

As part of the routine assessment discussed in section 4.1.1, core reactivity is calculated. Reactivity, ρ describes the deviation of an effective multiplication factor, k_{eff} from unity. It is defined by the equation

$$\rho = \frac{k_{\text{eff}} - 1}{k_{\text{eff}}}. \tag{4.7}$$

Mathematically, ρ is a dimensionless number. However, reactivity is often discussed in units of milliniles (mN) or percent mille (pcm), both of which are at the scale of 10^{-5}, as a means of communicating the generally very small magnitude of ρ. The unit nile was invented by the British nuclear industry and is still the primary unit used to communicate reactivity in the UK today. A value of $k_{\text{eff}} = 1$ represents a stable critical state of constant power. $k_{\text{eff}} > 1$ represents a supercritical state and $k_{\text{eff}} < 1$ a sub-critical state.

Reactivity is an important concept as it is a fundamental requirement to be able to shutdown a nuclear reactor and for it to remain sub-critical should a fault occur. The level of reactivity clearly impacts how easy this is to achieve. Shutdown and hold-down is managed by the insertion of the primary shutdown system (PSD) which normally consist of boronated stainless steel rods which act as strong neutron absorbers. The PSD is designed with significant redundancy to account for partial failure (in the unlikely event that some rods do not insert) or for routine maintenance of specific control rods which would render then unavailable to insert during the fault.

How far away from re-criticality the reactor is is often referred to as the sub-critical margin or shutdown margin (SDM). The SDM is a balance of the positive reactivity from the nuclear fuel, the negative reactivity from the control rods and the reactivity due to reactor 'poisons' such as ^{135}Xe and ^{149}Sm. Due to the half-lives and fission yields, reactor poisons are often simply taken to be ^{135}Xe as a first-order approximation. In SDM and reactivity balance assessments, the first two parts of this equation are effectively constant for a given core state. However, it should be recognized that the core state naturally loses reactivity as the nuclear fuel is burnt up. This leads to core reactivity decreasing with time throughout the operational run and SDM increasing. The contribution from reactor poisons however is not constant and is determined by the magnitude of the reactor transient, and evolves with time. As ^{135}Xe is primarily a daughter product of ^{135}I, consideration of fission yields and decay constants for these isotopes is necessary to calculate the contribution to overall reactivity and how this evolves with time. The pin inventory calculations described in section 4.1.2 are a key input in these assessments.

Xenon override is the term used to describe the condition where build-up of ^{135}Xe results in the subsequent decrease in reactivity that could lead to a reactor shutdown if there is insufficient positive reactivity to maintain a chain reaction. It also describes the converse condition—where the subsequent decay can lead to re-criticality should there be enough positive reactivity in the core and enough of the PSD has failed. Being unable to maintain a chain reaction would result in a significant commercial loss, but would not increase nuclear risk. In order to manage operational risk, reactor physicists at operational facilities often provide advice on whether a particular reactor transient may lead to an unplanned shutdown. The latest core state is used to provide the most current pin inventory assessment and the transient is then modelled to assess the effect on the ^{135}Xe inventory. Figure 4.3 illustrates how the reactivity contribution of ^{135}Xe evolves over time following power transients. It can be seen that following a drop in reactor power, the ^{135}Xe inventory initially increases before decreasing to its new equilibrium value. The converse is observed following an increase in reactor power. This is due to the delayed production of ^{135}Xe, with it primarily being a daughter product of ^{135}I. The resulting change in ^{135}Xe inventory will lead to a change in overall reactivity that the automatic control loops will attempt to compensate for by moving the control rods. Should the magnitude of the change in reactivity require the rods to move outside the range allowed by the LCOs, the reactor will either trip automatically or will be manually shutdown by the operator to maintain positive control and be within the

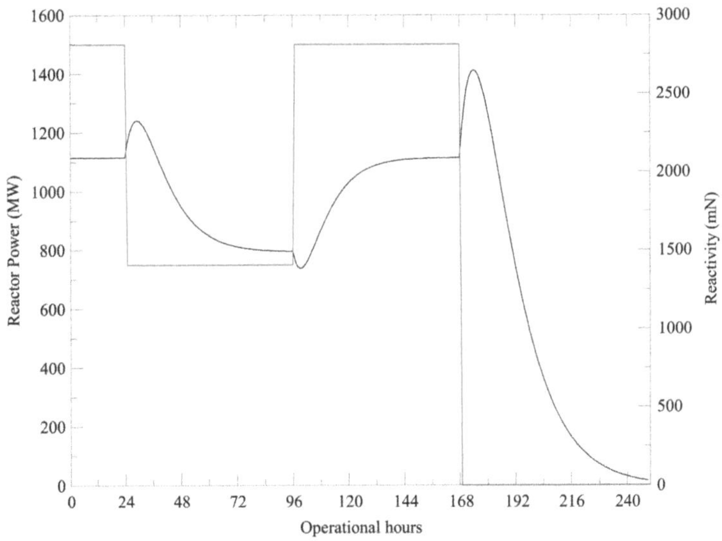

Figure 4.3. Typical reactivity worth due to ^{135}Xe during reactor power transients. Reactivity worth in milliniles (mN) of ^{135}Xe has been calculated during modelled transients for an AGR and is plotted in black. Reactor power, in megawatts (MW) is plotted in red. Although xenon reactivity worth will be dependent on specific fuel design, the behaviour will be representative for all reactor types using low-enrichment uranium as a fuel source.

requirements of the safety case. The reactivity 'worth' of the control rods is often non-linear due to the rod components having differing neutron absorption properties. The likelihood of the control rods being able to compensate for the transient and avoid an unplanned shutdown is therefore dependent on initial rod position within the core. Typically when rods are at the extremes of allowable positions, normally at the beginning and end of operational runs, the magnitude of travel required to compensate for the transient is greatest. It is at these points in the operational run that the risk that the control system cannot compensate to avoid an unplanned reactor shutdown is largest.

Although unplanned shutdowns are not desirable from a commercial standpoint, they do not increase the risk of a radiological release. What must be ensured is that the PSD system can terminate any fault that could lead to a radiological release under all credible conditions. To guarantee this, during core design, fuel loadings are chosen such that the negative reactivity of the PSD is greater than the reactivity of the core and the positive contribution to reactivity from ^{135}Xe which occurs several hours post-shutdown. In figure 4.3, it can be seen that initially following a reactor shutdown, SDM will increase as the ^{135}Xe inventory builds up. This is helpful for maintaining reactor shutdown. However, after around 8 h, the ^{135}Xe inventory will decrease and after 16 h will be less than at the point of shutdown. At this point the ^{135}Xe inventory is effectively reducing SDM. It is important to account for the contribution to reactivity of ^{135}Xe as if not properly considered, there is a risk that the reactor could go critical after initial shutdown, which would lead to fault escalation and potential radiological release.

4.1.5 Fuel performance and coolant activity monitoring

The safety case for an operating nuclear facility will identify all credible faults and provide appropriate LOPs to ensure the risk of radiological release is minimized and within the ALARP principle. In practice, this means that faults which have a return frequency of up to 10^{-4} require one LOP whilst those with 10^{-3} or greater require two LOPs. Faults with a return frequency lower than 10^{-6} are considered *incredible* and do not require a formal LOP to be claimed. Almost all reactor faults under consideration will ensure that the fault is terminated before significant damage to the primary fission product barrier (the fuel pin cladding) occurs, thus avoiding a large radiological release. This is achieved by controlling the start-of-fault fuel conditions. In practice, the primary coolant activity is monitored to ensure that in normal operation the activity levels are low enough to tolerate some fuel damage in a fault and still ensure that the radiological release is low enough to avoid harm to the public. The primary coolant activity is monitored using high-purity germanium (HPGe) gamma-spectroscopy systems. Gaseous fission products are measured to infer the condition of the fuel and confirm that overall activity levels are within safe operating levels. Figure 4.4 shows a typical spectrum obtained with the Gaseous Activity Monitoring (GAM) system at Hartlepool Power Station in the United Kingdom. It can be seen from this figure that, as one may expect, the sample has high intrinsic background. Background activity within the sample is from a number of sources. Isotopes such as ^{60}Co appear within the spectrum due to activation of steel structures within the core such as the fuel pins themselves. ^{41}Ar and ^{19}O are unique to the AGR and appear because of activation of dopant within the primary CO_2 coolant. Background levels of noble gas fission products such as ^{133}Xe and ^{87}Kr

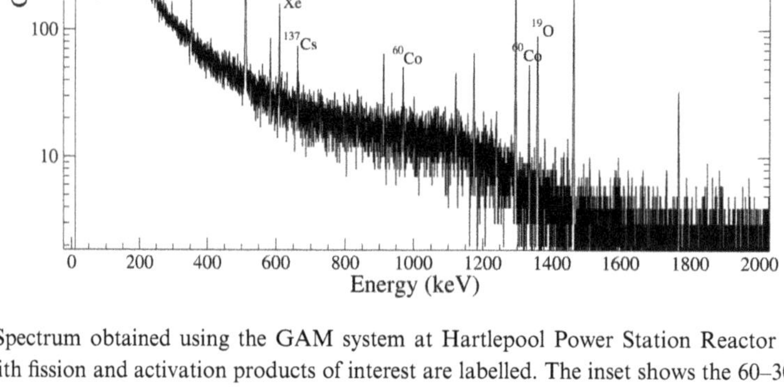

Figure 4.4. Spectrum obtained using the GAM system at Hartlepool Power Station Reactor 1. γ-ray lines associated with fission and activation products of interest are labelled. The inset shows the 60–300 keV region with measured xenon isotopes labelled.

are due to 'tramp' uranium—microscopic levels of uranium ceramic on the outside of the fuel pins left behind in the manufacturing process—undergoing fission and producing background levels of these isotopes within the primary coolant.

^{133}Xe is the primary isotope used to diagnose fuel damage and is compared to its theoretical yield to assess whether an excess amount is within the primary coolant and therefore indicative of fuel damage. Other short-lived isotopes such as ^{135}Xe, ^{87}Kr and ^{88}Kr are monitored as their isotopic ratios can give information on the damage mechanism and the overall potential inventory release. In normal operation, with the fuel pin cladding intact, the measured activity, $A(I)$, for a given isotope will be proportional to the product of its fission yield, $Y(I)$, the reactor power, P, and the mass of 'tramp' uranium on the pin surface, m. If the fuel pin cladding is breached, there may be an additional contribution to the measured activities from the pin inventories. This will be a function of decay constant, $\lambda(I)$ and will be dependent on the hold-up process delaying release into the primary coolant. This relationship can be written as

$$\frac{A(I)}{Y(I)\cdot P} \propto C\cdot\left(\frac{F}{\lambda(I)}+m\right), \tag{4.8}$$

where F is the fractional release from a fuel pin and C a constant the mass of primary coolant and neutron flux. Assuming that only 'tramp' uranium is contributing to the measured coolant activity, the ratio of activity $A(I)$ to the product of yield $Y(I)$ and power P for a given isotope should be independent of half-life or decay constant $\lambda(I)$. Therefore, when plotted against λ, this ratio should give a horizontal best-fit line which represents the mean ratio of activity to yield for a given reactor power. Isotopes in excess of this value will have a contribution to the activity due to fractional release from the fuel pin, indicating a fuel failure is present in the core. The magnitude of deviation can be used to calculate F and estimate the number of failed pins in the core. The gradient of the best-fit line may also be used to infer properties of the failure site, as larger breach sites will allow shorter-lived isotopes to diffuse into the primary coolant and enhance their measured activities. This information can be used to assess the likely damage mechanisms. It should be noted that equation (4.8) relies on the measured isotopes being in equilibrium concentrations and so is only applicable once the reactor has been operating at steady power for a number of half-lives of the longest-lived isotope.

4.2 Nuclear security and safeguards

Nuclear operators are duty-bound and legally required to ensure that the technology and materials specifically required for nuclear operations are tightly controlled. This ensures that the technology is used only for peaceful purposes and does not spread to actors or states intent on employing it for weapons production or other nefarious purposes. This section will discuss how nuclear data are used in the safeguarding of nuclear material and how reactor monitoring is being used to help distinguish peaceful operation from operations undertaken to aid weapons development.

4.2.1 Nuclear safeguards—tracking fissile material

The Treaty on the Non-Proliferation of Nuclear Weapons (NPT) [8] is a landmark treaty designed to prevent the spread of nuclear weapons technology. It also promotes the peaceful use of nuclear technology and aims to achieve complete disarmament. The treaty promotes cooperation between state parties and equal access to nuclear technology while safeguards prevent the diversion of fissile material for weapons use. Compliance with the treaty is verified through inspections conducted by the IAEA. An operating nuclear facility is obligated through its safeguards arrangements to ensure that its nuclear inventory is tracked at every stage and each gram of fissile material is accounted for. Nuclear Materials Balance figures, also known as Inventory Difference, reflect differences in the nuclear material accounts and inventories derived from measurements. Monthly Inventory Change (IC) reports are produced to report the changes in materials balances which have occurred due to operational activities such as fuel delivery and disposal, as well as the change in fissile material due to reactor operation. Facilities can be inspected at any time and must be able to account for all their fissile material inventory and be able to show where it is and where it has been. Of primary concern for the safeguards community is the amount of plutonium, uranium and thorium at a given facility. Although nuclear facilities within the UK do not employ plutonium or thorium as fuels, isotopes of these elements are produced as part of the fission process. The amounts produced are heavily dependent upon the operational histories of the reactors. Because it is not possible to directly measure the in-core inventories, the amounts must be inferred through calculation. The calculated values are then compared to independent assessments and are subject to verification once the fuel has been discharged. Reactor physics codes used to calculate the pin inventories for decay heat assessments discussed in section 4.1.3 are also employed to calculate the plutonium and thorium inventories. The complex production and decay mechanisms must be taken into account and therefore accurate nuclear data are essential to ensuring the estimates are within the required accuracies. The output from these assessments help demonstrate to the regulators, and ultimately the public, that the technology is controlled effectively and that the technology is being used solely for peaceful energy production, and not weapons development.

4.2.2 Monitoring for nuclear security—antineutrino monitoring

UK civil nuclear reactors evolved from reactors originally built to breed plutonium for weapons production. The UK AGR evolved from the UK MAGNOX design, which was originally built to assist in growing the UK's plutonium stockpile. Graphite-moderated reactors are well-equipped to produce plutonium in useful amounts for weapons production. It is known that the North Korean reactors at Nyongbyon are derived from a MAGNOX design and are being used for these purposes—an unintended consequence of not previously considering nuclear information security adequately. How these reactors are operate—their operational histories—have a significant effect on the plutonium production and the overall inventory. As the reactor fuel is irradiated over time, the plutonium produced through the fission of uranium is 'burnt up'—it itself undergoes neutron capture, fission and decay. A tell-tale sign that a reactor is being used to breed plutonium are frequent refuelling and short periods of operation. In these cases, the fuel is being irradiated for short periods to produce the plutonium and then discharged before a significant amount of the plutonium is 'burnt up' and is subsequently harvested for weapons development. When a state develops nuclear technology, it is therefore important that the international community can verify that it is being operated only for peaceful purposes. In more open states, this is done through agreed inspection regimes, as outlined in section 4.2.1. In less open states, agreement for international bodies to inspect the facilities can be more difficult to obtain. There is a significant amount of interest in developing techniques to remotely monitor the operation of nuclear facilities so that the international community can independently verify the use of the facilities. One such method that is being researched is the detection of antineutrinos to confirm the operational mode of a nuclear reactor.

Nuclear reactors are one of the strongest sources of electron antineutrinos on Earth. Electron antineutrinos from reactors originate from the β^- decay of various fission product daughters (see equation (1.13)). Each daughter will, on average, decay three times and so produce three antineutrinos; this leads to approximately six antineutrinos being produced per fission event. The total number of antineutrinos per fission event does not vary significantly with fissile isotope. However, the population of daughter products does, resulting in substantial differences in the emitted antineutrino energy spectrum. The possibility of using antineutrinos for safeguards applications was discussed as long ago as 1978 when it was recognized that the dynamics of ^{239}Pu could be inferred by measuring the antineutrino spectra from operational reactors. It was also recognized that the spectral evolution could be used to infer reactor properties such as fuel burn-up and time-averaged power [9]. More recently, it has been proposed that an antineutrino spectrum for a reactor can be represented by the spectra from each nuclide that has fissioned [10]. The antineutrino flux will be dependent upon the fission product inventories, which, in turn, will depend upon the operating mode and history. If the reactor design is known and modelled, the theoretical antineutrino flux can be calculated for various operating modes. The main modes of interest would be full power operation, part-power and shutdown. A detector capable of detecting antineutrinos in quantities large enough to produce measured flux profiles would theoretically be able to distinguish between modes of operation and infer what the facility is doing at any point in time. If a facility was seen to be operating for short periods and frequently shutting down, this would alert suspicions since most electricity generators plan to operate at full load for as long as possible to maximize output. As previously mentioned, short periods of operation favour plutonium production and so could be a sign that the facility is operating for purposes other than peaceful energy production. Obtaining an antineutrino spectrum across energy such as that shown in figure 4.5 over a period of time could give evidence as to how a facility is being operated and be used to verify the information which has been declared to the safeguards inspectors.

Neutrinos by their nature barely interact with matter and can travel large distances without impediment. A detector could therefore in theory be placed many kilometres

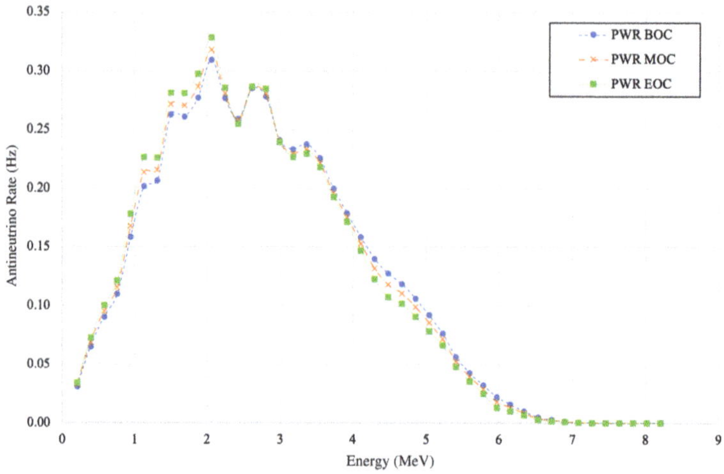

Figure 4.5. The spectral result of antineutrino rates against energy (MeV) at three points in the reactor fuel cycle (days 1, 800, and 1600) for the SONGS reactor unit. (Reproduced with permission from [11]. Copyright 2024 the American Physical Society.)

away and remotely or covertly undertake monitoring. Antineutrinos are detected via the inverse-beta-decay interaction in a medium such as gadolinium-doped water. Here, antineutrinos interact with a proton in the detector medium to produce an position–neutron pair:

$$\bar{\nu}_e + \mathrm{p} \rightarrow \mathrm{e}^+ + \mathrm{n}. \tag{4.9}$$

The positron is detected by the Cherenkov radiation emitted as it exceeds the speed of light in the medium. The neutron produced in this process will ultimately be captured by the dopant in the medium, after which Cherenkov radiation is emitted during Compton scattering. The expected antineutrino flux $N(E_{\bar{\nu}_e})$ measured by a detector at distance L over a time period, T, can be given by

$$N(E_{\bar{\nu}_e}, L) = \frac{n_p T}{4\pi L^2} \sum_l N_l^f \phi_l(E_{\bar{\nu}_e}) \sigma(E_{\bar{\nu}_e}) P_{ee}(E_{\bar{\nu}_e}, L), \tag{4.10}$$

where n_p is the number of free protons. The contribution to the measured antineutrino flux from a given isotope, l is represented by the fissile fraction N_l^f and its unique spectrum, ϕ_l. Scaling factors for expected losses due to antineutrino oscillations and inverse-beta-decay cross sections are represented by P_{ee} and σ.

There have been prototype detectors built and deployed close to operating reactors as proof of concept. These detectors are typically a few cubic metres in volume and are deployed tens of metres from the active core. Differing technologies have been trialled. For example, the PANDA project reports results from a plastic scintillator-based detector deployed 36 metres away from the 3.4 $\mathrm{GW_{th}}$ core of Ohi Power Station PWR [12]. This work showed that it is, in principle, possible to detect the on/off state of the reactor by measuring the antineutrino flux. Liquid scintillator detectors have been deployed at the San Onofre Nuclear Generating Station (SONGS) facility. The detector was placed approximately 25 metres from the active core of the 3.4 GW_{th} PWR. Results reported show that this detector type can detect prompt reactor shutdown and monitor thermal power [13] and observe long-term changes in antineutrino flux due to the evolution of isotopic inventories [14].

In order to overcome the inherently-reduced detection efficiency, small prototype reactors need to be placed very close to the active core to have a reasonable chance of collecting enough data to make meaningful conclusions. In practice, it is unlikely that many operational facilities will tolerate a detector being placed so close to the active core. This will be particular true for states less inclined to cooperate with IAEA safeguards requirements. For practicable safeguards applications, larger detectors will need to be deployed so that antineutrinos may be detected at greater distances, potentially kilometres away from the active core of interest. There have been a number of large-scale high-profile experiments aimed at measuring reactor antineutrinos from extended distances. Though they have potential safeguards applications, the main object of these experiments was to investigate neutrino oscillation and mixing. The Kamioka Liquid Scintillator Anti-Neutrino Detector (KamLAND) collaboration have reported observations of antineutrinos from distant power reactors [15]. The KamLAND detector consists of 1 kton of ultra-pure liquid scintillator contained within a 13 m diameter spherical balloon made of transparent nylon, and is situated deep underground to reduce background from cosmic muons. The collaboration report detections from power reactors with distances ranging from 88 km to 295 km from the detector site. Recently, the SNO+ collaboration reported the first evidence of reactor antineutrinos in a Cherenkov detector [16]. SNO+ is located 2 km underground in an active mine, near Sudbury, Ontario. The detector consists of a 12 m diameter spherical acrylic vessel submerged in ultra-pure water, surrounded by an array of photo-multiplier tubes (PMTs). The nearest reactor was some 240 km away from the detector site. These collaborations demonstrate the potential for extreme remote detection, although further development will be required if they are to undertake potential safeguards applications.

The Double Chooz experiment reported detections of antineutrinos from the two-reactor Chooz Nuclear Power Plant [17]. The detector was located just over 1 km from the two 4.25 GW_{th} reactors at the Chooz facility underneath 300 MWE of rock overburden. Reactor data provided by the operator EDF allowed for more accurate simulation of the four main fissioning isotopes, ^{235}U, ^{239}Pu, ^{238}U, and ^{241}Pu, and therefore the predicted antineutrino flux. Measurements were in line with expectations and ruled out a no-oscillation hypothesis at the 95% confidence level. The results show that remote detection of antineutrinos from operational reactors can be achieved at distance and that the measured flux can be compared to the theoretical profile if initial reactor data and operational histories are known. Operational data from the Hartlepool nuclear reactor facility in the UK has been used to calculate an antineutrino flux profile for this specific reactor [11]. The results of these simulations calculating expected antineutrinos per fission event were compared to the SONGS PWR reactor and the Sizewell B PWR reactor. Due to the more frequent refuelling of the Hartlepool AGR reactor design, the number of antineutrinos per fission was much more consistent across the same time period. The PWR reactors showed a larger steady decline over their longer operating windows. It was also notable that the spectrum of antineutrino rate across energy for the AGR showed little to no variation across its fuel cycle. By comparison, there was a notable difference in the PWR spectra during the cycle. The variation in the spectrum of antineutrino rate across energy could therefore potentially be used to infer refuelling patterns and be used to verify the declared operations of a nuclear facility for safeguards purposes.

4.2.3 Monitoring for nuclear security—xenon emissions

Testing for the purpose of nuclear weapons development is prohibited under the Comprehensive Nuclear Test Ban Treaty (CTBT) [18]. As of March 2025, 187 states have signed or ratified this treaty. In order to enforce the treaty, the International Monitoring System (IMS) has been developed. This system is a unique global network that utilizes four complementary technologies to look for signs of nuclear weapons testing. These technologies are: seismic monitoring stations, hydroacoustic monitoring stations, infrasound monitoring stations and radionuclide monitoring stations. It is the latter technology that is of interest here and will be discussed further. It is designed to provide a 90% detection probability for a 1 kT TNT-equivalent atmospheric nuclear explosion within 14 days of release [19]. A detailed review of the radionuclide sub-network can be found in [20]. The radionuclide stations employ noble gas monitoring systems that look primarily to detect radio-xenon species from atmospheric samples. ^{133}Xe, ^{135}Xe, ^{133m}Xe and ^{131m}Xe are considered the most useful isotopes as they are not naturally occurring, are produced in nuclear fission reactions and have reasonably long half-lives to allow for detection days to weeks after initial production. The global IMS network looks for increases in the radioxenon background and uses atmospheric transport modelling to calculate the likely source term location. The IMS system successfully detected the 2013 nuclear test announced by the Democratic People's Republic of Korea (DPRK). Isotopic ratios of ^{133}Xe and ^{131m}Xe were consistent with a nuclear explosion event occurring more than 50 days prior to detection, which was consistent with the test date announced by the DPRK. Three-dimensional atmospheric transport modelling (ATM) was used to calculate the travel path of the radionuclides to estimate a potential test site within North Korea.

A method has been proposed [21] where ^{135}Xe/^{133}Xe and ^{133m}Xe/^{131m}Xe ratios can be used to assess whether the source term is likely to be from a civil facility or from a nuclear explosion, and therefore give evidence as to whether an unsanctioned nuclear test has taken place. Often referred to as a Kalinowski plot, an example is given in figure 4.6 using data from measurements at the Hartlepool nuclear power station. Data were obtained during 'blowdown' periods which occur when the reactor is shutdown for refuelling outages. The primary CO_2 coolant of the AGR is discharged through a filtered route before a fraction of the fuel within the core is replaced. Within these discharges, some radioxenon is released, in amounts well below levels that would have any impact to the public or environment. Data points

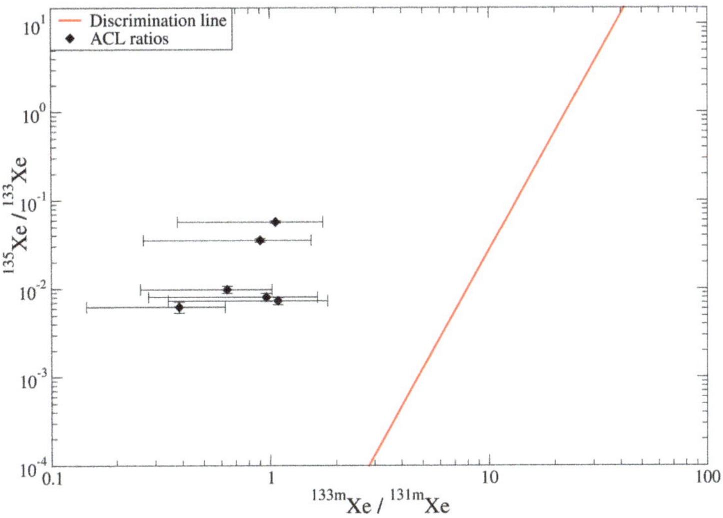

Figure 4.6. $^{135}Xe/^{133}Xe$ and $^{133m}Xe/^{131m}Xe$ ratios obtained using the data from measurements of Maypack units *in situ* at the Active Chemical Laboratory (ACL) in Hartlepool power station. The 4-isotope 'discrimination' line from [21] is plotted. All calculated points lie to the left of this line, within the 'civilian' region. (Reproduced from [22]. CC BY 4.0.)

which lie to the right of the 'discrimination' line would have an excess of isomeric xenon. This would be interpreted as the source being from a nuclear explosion. The data points from the Hartlepool reactor lie to the left of the line and are correctly classified as being from a civil nuclear source. The 'discrimination' lines defined in [21] have been derived from publicly available data of emissions from civil nuclear facilities, theoretical calculations and the limited data available from previous nuclear tests.

Knowledge of the global xenon background has been derived from the publicly available data from commercial and research reactors, as well as medical isotope production facilities (MIPs) [23–25]. These facilities are considered to be the main sources of radioxenon within the atmosphere. Due to radioxenon isotopes having almost negligible impact on both public and environmental safety, regulating bodies often do not impose strict reporting criteria for emissions of these isotopes, instead, often only requiring the total noble gas inventory to be reported. The data used in [23–25] are therefore somewhat coarse and potentially over-simplified. A comprehensive overview of the challenges faced in this area is given in [26]. In order for the discrimination method proposed in [21] to be reliable, background radioxenon levels must be well understood. To fully understand the radioxenon background, the emissions from peaceful, civil nuclear facilities must also be well understood so as these can be easily distinguished and accounted for within IMS detections. To better understand emissions from these facilities the Source Term Analysis of Xenon (STAX) [27] project has been installing gamma-spectroscopy systems to measure stack emissions from partnering civil nuclear facilities, the aim being to directly measure emissions at source, to better characterize the contribution to the global

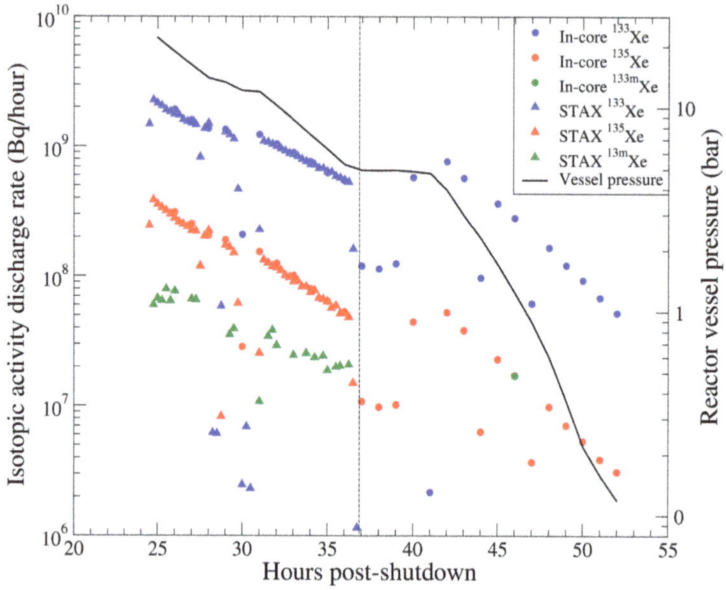

Figure 4.7. Discharged Xe activity in Bq/hour during a reactor blowdown at Hartlepool nuclear power station prior to planned refuelling outages. Figures show release activity measured using the installed STAX system at the exhaust stack and calculated discharges based on in-core measurements using the GAM system. (Reproduced from [22]. CC BY 4.0.)

xenon background. One such monitor has been installed at the Hartlepool nuclear facility, data from which have been used in figure 4.7. Obtaining isotopic 'fingerprints' of civil nuclear facilities will help to greatly improve knowledge of the radioxenon background and emissions from civil facilities. The Hartlepool nuclear facility in the UK has been working with the international monitoring community to help characterize the emission from civil nuclear facilities to aid these efforts [28]. They have provided direct measurements of radioxenon releases to help characterize the profile from a civil nuclear facility. An example of a radioxenon release profile is given in figure 4.7. This figure shows the radioxenon release that occurs during a 'blowdown' at an AGR. Data were obtained at the Hartlepool nuclear power plant using the in-core coolant activity monitoring system and a STAX monitor installed at discharge stack. Data obtained in this campaign showed that the ^{133}Xe annual inventory release was an order of magnitude less than that assumed in [23] and showed that the release profile for an AGR was not a continuous, but was periodic. This has led to better understanding of facility release, although more work is required to characterize other facilities of differing reactor design.

References

[1] Askew J R, Fayers F J and Kemshell P B 1966 General description of the lattice code WIMS *J. Brit. Nucl. Energy Soc.* **5** 564–85

[2] Rhodes J, Smith K and Lee D 2006 *CASMO-5 Development and Applications* (American Nuclear Society)

[3] Chadwick M B *et al* 2011 ENDF/B-VII.1 nuclear data for science and technology: cross sections, covariances, fission product yields and decay data *Nucl. Data Sheets* **112** 2887–996
[4] The JEF-2.2 Nuclear Data Library OECD *Nuclear Energy Agency* https://www.oecd-nea.org/jcms/pl_36783/the-jef-2-2-nuclear-data-library
[5] Nakagawa T *et al* 1995 Japanese evaluated nuclear data library version 3 revision-2: JENDL-3.2 *J. Nucl. Sci. Technol.* **32** 1259–71
[6] Hutt P K, Gaines N, McEllin M, White R J and Halsall M J 1991 The UK core performance code package *Nucl. Energy* **30** 291–8
[7] Burstall R FUKAEA 1979 *FISPIN—A Computer Code for Nuclide Inventory Calculations, Technical Reprt* Risley Nuclear Power Development Establishment
[8] UNTS 1968 *Treaty on the Non-Proliferation of Nuclear Weapons* Treaty 10485, vol 729 p 161 United Nations
[9] Borovoi A A and Mikaélyan L A 1978 Possibilities of the practical use of neutrinos *Sov. At. Energy* **44** 589–92
[10] Huber P 2011 Determination of antineutrino spectra from nuclear reactors *Phys. Rev.* C **84** 024617
[11] Bogetic S, Mills R, Bernstein A, Coleman J, Morgan A and Petts A 2024 Antineutrino flux from the EDF Hartlepool *Phys. Rev. Appl.* **21** 064051
[12] Oguri S, Kuroda Y, Kato Y, Nakata R, Inoue Y, Ito C and Minowa M 2014 Reactor antineutrino monitoring with a plastic scintillator array as a new safeguards method *Nucl. Instrum. Methods Phys. Res.* A **757** 33–9
[13] Bernstein A, Bowden N S, Misner A and Palmer T 2008 Monitoring the thermal power of nuclear reactors with a prototype cubic meter antineutrino detector *J. Appl. Phys.* **103** 074905
[14] Bowden N S, Bernstein A, Dazeley S, Svoboda R, Misner A and Palmer T 2009 Observation of the isotopic evolution of pressurized water reactor fuel using an antineutrino detector *J. Appl. Phys.* **105** 064902
[15] Eguchi K *et al* 2003 First results from KamLAND: evidence for reactor antineutrino disappearance *Phys. Rev. Lett.* **90** 021802
[16] Allega A *et al* 2023 Evidence of antineutrinos from distant reactors using pure water at SNO+ *Phys. Rev. Lett.* **130** 091801
[17] Abe Y *et al* 2012 Indication of reactor $\bar{\nu}_e$ disappearance in the Double Chooz Experiment *Phys. Rev. Lett.* **108** 131801
[18] UN 1996 *Comprehensive Nuclear-Test-Ban Treaty, United Nations Resolution* A/RES/50/245, document A/50/927 United Nations
[19] Medici F 2001 IMS radionuclide network of the CTBT *Radiat. Phys. Chem.* **61** 689–90
[20] Goodwin M A, Davies A V, Britton R, Miley H S, Eslinger P W, Hoffman I, Ungar K, Mekarski P and Botti A 2024 Radionuclide measurements of the International Monitoring System *J. Environ. Radioact.* **272** 107357
[21] Kalinowski M *et al* 2010 Discrimination of nuclear explosions against civilian sources based on atmospheric xenon isotopic activity ratios *Pure Appl. Geophys.* **167** 517–39
[22] Petts A, Bowyer T, Friese J, Goodwin M and Milbrath B 2024 Measurements of radioxenon activities during periods of gaseous release from an Advanced Gas-cooled Reactor *Phys. Rev. Appl.* **22** 044060
[23] Kalinowski M B and Tuma M P 2009 Global radioxenon emission inventory based on nuclear power reactor reports *J. Environ. Radioact.* **100** 58–70

[24] Kalinowski M B, Tayyebi P, Lechermann M and Tatlisu H 2021 Global radioxenon emission inventory from nuclear research reactors *Pure Appl. Geophys.* **178** 2711–39

[25] Hoffman I and Berg R 2018 Medical isotope production, research reactors and their contribution to the global xenon background *J. Radioanal. Nucl. Chem.* 165–73

[26] Bowyer T W 2021 A review of global radioxenon background research and issues *Pure Appl. Geophys.* **178** 2665–75

[27] Metz L, Bowyer T, Burnett J, Dion M, Eslinger P, Friese J, Doll C, McIntyre J and Schrom B 2022 Source Term Analysis of Xenon (STAX): an effort focused on differentiating man-made isotope production from nuclear explosions via stack monitoring *J. Environ. Radioact.* **255** 107037

[28] Goodwin M A *et al* 2024 Characterising the radionuclide fingerprint of an advanced gas-cooled nuclear power reactor *Pure Appl. Geophys.* https://doi.org/10.1007/s00024-024-03488-y

Chapter 5

Long-lived radioactive waste: nuclear data requirements, issues for safe storage and potential for transmutation

David Jenkins

Fission reactors generate long-lived nuclear waste which will need safe storage long into the future. This category of waste has two principal origins: long-lived fission products and the so-called minor actinides including plutonium. Understanding the extent to which these are produced in a given reactor design requires detailed understanding of the relevant nuclear reactions which produce them. Here, we explain the relevance to this problem of neutron-induced reaction studies exploring how to ensure that the data obtained have sufficient precision and accuracy. In particular, we will focus on the value of neutron time-of-flight systems such as the nToF facility at CERN which efficiently allow such reaction studies to be carried out over a wide energy range. Latterly, we will look at issues in safe storage of long-lived waste focussing on the issue of neutrons produced through (α,n) reactions involving minor actinides in waste. Finally, we will briefly discuss nuclear transmutation which has often been touted as a more permanent solution to long-lived nuclear waste but which presents a range of practical and financial challenges.

Concepts: Long-lived waste, long-lived fission products, minor actinides, neutron-induced reactions, neutron time-of-flight facilities,

5.1 Long-lived radioactive waste

An important issue in terms of the legacy of nuclear fission energy is the extent to which it leads to long-lived radioactive waste. The generation of such waste complicates the handling of spent fuel and implies long terms liabilities in the handling of such materials. The nuclear data required to understand the production of long-lived waste are well constrained for conventional fission reactors, although there are continual efforts to improve such data not only to more precisely constrain the relevant nuclear reactions but also to remove engineering uncertainties in the

doi:10.1088/978-0-7503-5102-7ch5 © IOP Publishing Ltd 2025. All rights, including for text and data mining (TDM), artificial intelligence (AI) training, and similar technologies, are reserved.

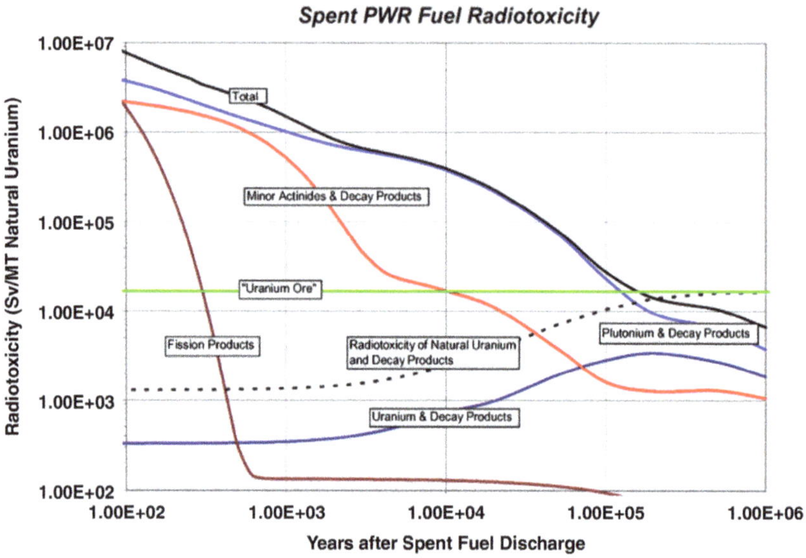

Figure 5.1. Radiotoxicity of components in spent nuclear fuel from a typical PWR reactor in units of sieverts per metric ton of natural uranium. The contribution of the different components may vary depending on the design of the reactor under consideration. The horizontal line is the radiotoxicity represented by uranium ore, in other words, the activity of the uranium in its natural form before being processed and used as fuel within a reactor. (Reproduced with permission from [1]. Copyright 2011 Elsevier.)

safe operation of reactors. When we consider the future of the nuclear industry where different reactor types are foreseen, potentially employing fast neutrons or different fuel types, additional nuclear data will undoubtedly be needed to understand more fully the waste inventory likely to be produced. One should note the strong parallels with current efforts in relation to activation and waste production in future fusion reactors as discussed in chapter 2.

Here, we begin by focussing on the origin and composition of long-lived radioactive waste from a conventional fission reactor. Figure 5.1 illustrates the radiotoxicity of a typical PWR reactor over time: the most significant contributors being *plutonium*, *minor actinides* and *fission products*. In the next sections, we discuss the origin of these different components.

5.1.1 Plutonium and minor actinides

In a conventional reactor, uranium fuel is typically enriched to 4%–5% in ^{235}U while the remainder of the uranium is in the form of ^{238}U. The ^{238}U does not play a direct role in the fission process but its presence does strongly contribute to the build-up of the radioactive isotope inventory within the fuel. Neutron capture on ^{238}U produces ^{239}U which successively beta decays to ^{239}Pu—a process known as *fertile conversion*. ^{239}Pu is a fissile isotope so its presence will also contribute at some level to energy generation within the reactor, as well as generating fission products which follow a slightly different mass distribution to those produced in fission of ^{235}U.

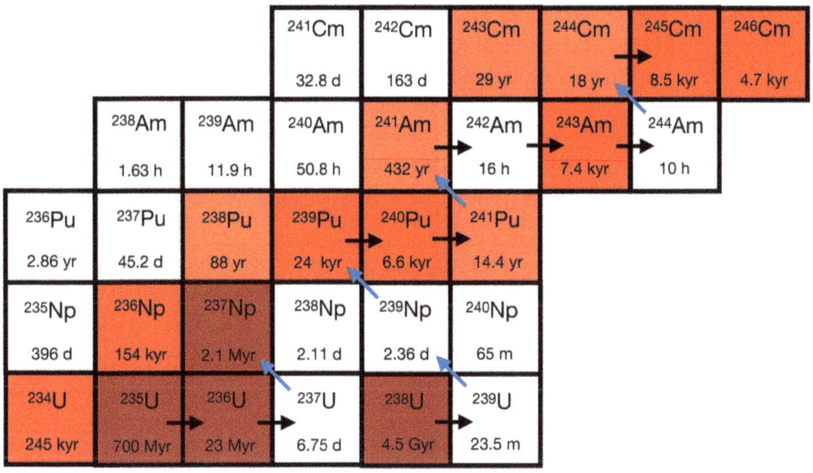

Figure 5.2. A portion of the table of isotopes relevant to the production of minor actinides in a conventional fission reactor. Isotopes are labelled along with the half-life of the respective ground state. Long-lived species are shown in red, with the depth of the colour corresponding to the range of half-life. Some of the principal pathways are shown for successive neutron capture (black arrows) followed by beta decay (blue arrows) starting from ^{235}U and ^{238}U components in the fuel.

A secondary process of successive neutron capture on ^{239}Pu produces $^{240,\,241}$Pu; successive beta decay followed by neutron capture produces the minor actinides such as ^{241}Am and ^{244}Cm (see figure 5.2)[1]. It is the presence of plutonium and the minor actinides in spent fuel which generates most of the long-lived radioactive waste and, hence, the constraints on storage and disposal of such material (see figure 5.1). A precise knowledge of (n,γ) capture and (n,f) reactions will be of high value in calculating the inventory of minor actinides produced during the operation of a reactor.

5.1.2 Long-lived fission products

We already discussed in chapters 3 and 4 how fission products are responsible for the phenomenon of decay heat—something which has to be adequately accounted for to ensure safe reactor operation. A wide range of fission products are created in the fission of ^{235}U (see figure 1.18) but the majority are short-lived and decay away rapidly in spent nuclear fuel. Some, however, have half lives in the range of years/decades and persist in the environment following nuclear accidents or unintentional releases. A well-known example of this is the isotope ^{137}Cs which has a half-life of 30 years. The 662 keV gamma-ray emitted in the decay of this isotope is readily detected in soil samples across the North of England and Scotland many years after debris was deposited there from the Chernobyl disaster in 1986.

[1] Neutron capture on ^{238}U is not the only source of minor actinides since successive neutron capture on ^{235}U produces ^{237}U which beta decays into ^{237}Np which has a 2.1 Myr half-life (see figure 5.1).

Table 5.1. Seven LLFPs tabulated in ascending order of mass. The Q-values presented are those between the ground state of the parent and the ground state of the daughter. In some cases, this is not the favoured decay branch and decay goes preferentially to an excited state in the daughter. For convenience, half lives and Q-values are presented to either two or three significant figures (depending on the precision with which they are known) and without uncertainties. Relevant data are extracted from the National Nuclear Data Center (NNDC).

Isotope	Half-life (years)	$Q(\beta^-)$ (keV)
^{79}Se	3.26×10^5	151
^{93}Zr	1.61×10^6	91
^{99}Tc	2.11×10^5	298
^{107}Pd	6.5×10^6	34
^{126}Sn	2.18×10^5	378
^{129}I	1.57×10^7	189
^{135}Cs	2.3×10^6	269

More concerning from the point of view of nuclear waste are the subset of fission products that are very long-lived—some with half lives of order millions of years which will necessitate the safe storage of waste containing such fission products for indefinite periods. From a nuclear structure point of view, there is a relatively simple explanation for the occurrence of these long-lived fission products (LLFPs), namely that their Q-value for β decay is very low—typically a few hundred keV or less (see table 5.1). In a broader perspective, the accident of these low Q-values relates to the interplay of nuclear forces that defines the relative stability of isotopes.

5.1.3 Waste from future reaction designs

Our discussion on nuclear waste has so far focussed only on conventional fission reactors where the required nuclear data are relatively well known. This is not necessarily the case for the next generation of nuclear reactors known as Generation IV [3], which may be deployed as so-called *small modular reactors*. Indeed, there are many different reactor designs currently under consideration including fast reactors and molten salt reactors which may operate under quite different conditions and higher temperatures as compared to standard fission reactors leading to a different waste inventory [4]. In particular, for fast reactors, the higher neutron energies may influence the fission product mass distribution; for example, see this video of how fission yields of ^{239}Pu vary as a function of incident neutron energy. In general, where neutron energies are higher, the 'gap' between the low and high mass partitions of the fission fragment mass distribution tends to be filled in (see examples of this for fission of ^{232}Th and ^{233}U in figure 5.3).

There has been consistently strong interest worldwide in the potential offered by thorium-based reactors [5]. The proposed thorium fuel cycle is a little different from the uranium one since almost all thorium is in the form of ^{232}Th which is not a fissile

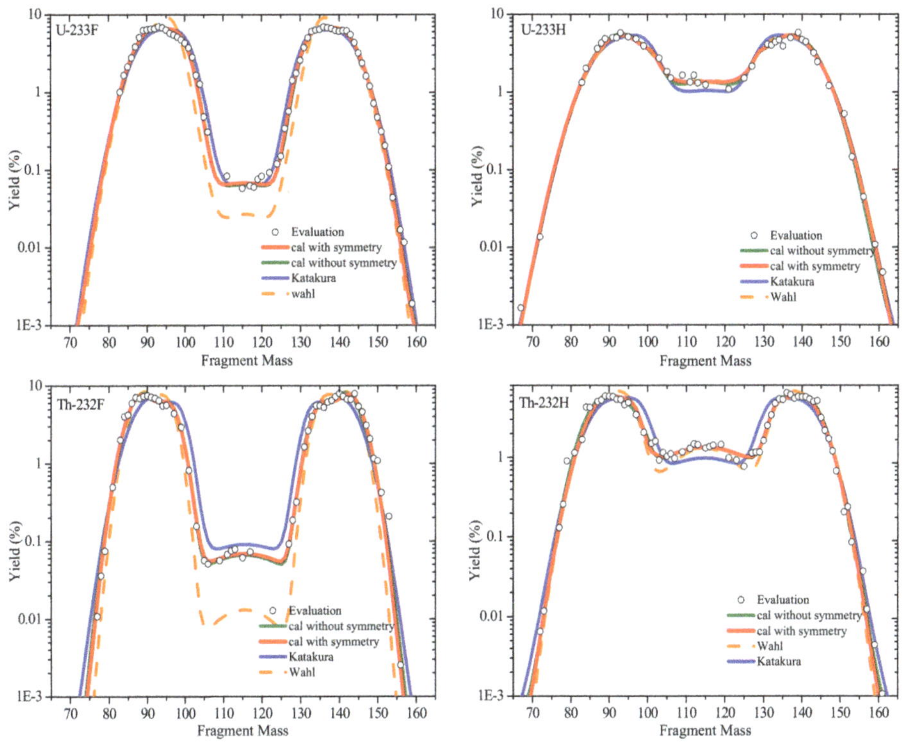

Figure 5.3. Fission yields for ^{233}U (top) and ^{232}Th (bottom) as function of mass. The evaluated data are compared with various theoretical models (see [7] for full details). The left-hand plots show data for fission yields using fission spectrum neutrons while the right-hand plots are for 14 MeV neutrons. (Reproduced with permission from [7]. Copyright 2017 Elsevier.)

isotope; it is necessary to capture neutrons onto ^{232}Th to transform it into the fissile ^{233}U which can be the basis of a reactor cycle. Nevertheless, there are certain advantages proposed for a thorium fuel cycle, not least that thorium is more abundant than uranium in the Earth's crust as well as the fact that the fuel cycle is more proliferation resistant. Unsurprisingly, given that thorium-based reactors are not currently operated, there is considerably less nuclear data available on the respective reaction pathways compared to those for conventional fission reactors [6]. Examples of fast fission data for ^{232}Th and ^{233}U are shown in figure 5.3.

5.1.4 Other contributions to waste

For completeness, one should note that nuclear fuels inevitably contain many other components such as fuel cladding made of different materials while the reactor itself may employ neutron moderators and other structural materials of different types. Neutron capture on these materials will also produce an inventory of radioactive material with varying hazards depending on their half-life such as ^{60}Co from neutron activation of ^{59}Co present in steel and ^{14}C from neutron capture on ^{13}C in graphite

[8]. There are clear parallels here with the activation of fusion tokamak components discussed in section 2.5.1.

Exercises

5.1. Working with figure 5.2 as a basis, try to generate a similar network of reactions and decays starting with ^{232}Th as the core fuel. Consider how the range of minor actinides produced might differ from the case of uranium fuel.

5.2. How might the long-lived fission products produced in fission based on thorium fuel differ from those involving uranium fuel? Comparing figure 5.3 with figure 1.18 may give a high level view. Further research into the available literature may provide a more detailed answer.

5.3. Thorium fuel is often stated to have a lower risk in terms of nuclear proliferation than standard fuels. Explore in the literature as to why this might be the case. What nuclear structure aspects are relevant here? How could improved nuclear data help us to constrain the risk still further?

5.2 Nuclear data needs to improve our understanding of long-lived waste production

As discussed in the previous sections, the main contributors to nuclear waste from conventional fission reactors are long-lived fission products and the minor actinides. Understanding their production and destruction will, in principle, require knowledge of a network of neutron capture and neutron-induced fission reactions[2]. In practice, much of the potential network of reactions is either not particularly important or the relevant data are already constrained with sufficient precision. How then should someone wanting to improve the nuclear data situation approach the problem? The task is not as difficult as it might seem as nuclear data needs are extensively reviewed on an ongoing basis by the international community—not only for the particular issues of waste generation discussed here but for other topics such as nuclear safeguarding and non-proliferation. Critical data needs are established on a consensus from international experts and are summarized in the Nuclear Energy Agency (NEA) High Priority Request List (HPRL) for Nuclear Data. Nearly all the data needs currently on the list (as of June 2025) involve neutron-induced reaction data, principally, neutron-induced fission on minor actinides as well as neutron capture, and neutron elastic/inelastic scattering on various isotopes. Let us therefore focus on how to achieve improved precision in neutron-induced cross sections. As was noted in chapter 1, there is a significant difference between neutron- and charged-particle induced reactions; this relates to the fact that neutrons are uncharged and therefore do not interact electromagnetically with the nucleus so do not experience Coulomb repulsion in the way that protons, alpha particles or other heavy ions would. As a consequence, neutron capture cross sections tend to

[2] Excellent resources for inspecting neutron-induced cross sections as a function of energy over a wide energy range include the JENDL-5 database and the EXFOR database [9].

increase to the lowest energies, the opposite of charged-particle induced reactions. Indeed, neutron-induced reactions are relevant over a very wide range of energy from thermal energies up to fast neutron energies. Neutron-induced reaction cross sections can be measured in specific energy ranges, for example, using thermal neutrons from a reactor or quasimonoenergetic neutron beams from some accelerator based system; however, in general, measuring energy-dependent cross sections for (n,γ) and (n,f) reactions is most efficient where a high flux of neutrons is incident on a target with a broad sweep of energies. This is readily achieved in a neutron time-of-flight facility where neutrons are generated with energies from GeV down to thermal from spallation reactions of high energy protons. Here, we discuss the Neutron Time-of-Flight (nToF) facility at CERN as an exemplar but there are other similar facilities worldwide such as GELINA in Belgium and LANSCE in the US.

Exercises

5.4. Open the current Nuclear Energy Agency (NEA) High Priority Request List (HPRL) for Nuclear Data. Choose a selection of the requests and consider what would be the best route to obtaining such experimental data. What challenges might be encountered—access to materials, systematic uncertainties, access to relevant apparatus?

5.2.1 The nToF facility at CERN

CERN is most commonly associated with the Large Hadron Collider whose discovery science is focussed on the most fundamental level of matter. This is not, however, the only active experiment at CERN and, indeed, there are two important nuclear physics facilities present on site: the ISOLDE facility which produces exotic nuclei for studies of nuclear structure such as measurements of nuclear masses and nuclear beta decay[3] and the nToF facility [10] which carries out nuclear data measurements on neutron-induced reactions relevant to both industrial applications such as fission reactors and nuclear astrophysics. When the CERN accelerator complex is operating, these two facilities often require the majority of the protons being accelerated at any given time. Indeed, the facilities take protons at earlier stages of the accelerator chain: 1.4 GeV protons from the PS Booster accelerator in the case of ISOLDE and 20 GeV protons from the PS accelerator in the case of nToF. The 20 GeV protons are incident on a lead spallation target (see figure 5.4) where they give rise to a high flux of neutrons across a broad range of energies from GeV down to thermal energies. Each incident proton can produce hundreds of secondary neutrons via spallation. Importantly, the 20 GeV proton beam is highly pulsed which means that over an appreciable flight distance, neutrons are separable by energy through their corresponding time of flight. The longer the flight distance, the higher the precision in neutron energy achievable—we quantify this in more detail below.

[3] ISOLDE is an important facility for study of the beta decay of radioactive nuclei relevant to reactor decay heat discussed in chapter 3. Its relevance to production of medical isotopes is also discussed in chapter 6.

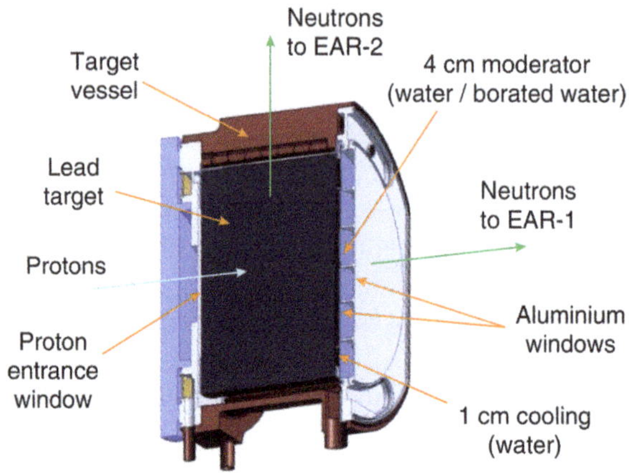

Figure 5.4. Cross-sectional view of the nToF spallation system showing the lead target and water cooling. The 20 GeV proton beam enters from the left. The neutron beam to EAR1 exits to the right and the beam to EAR2 exits vertically. The neutrons produced pass through a 4 cm thick moderator of water or borated water. (Reproduced from [11]. CC-BY-4.0.)

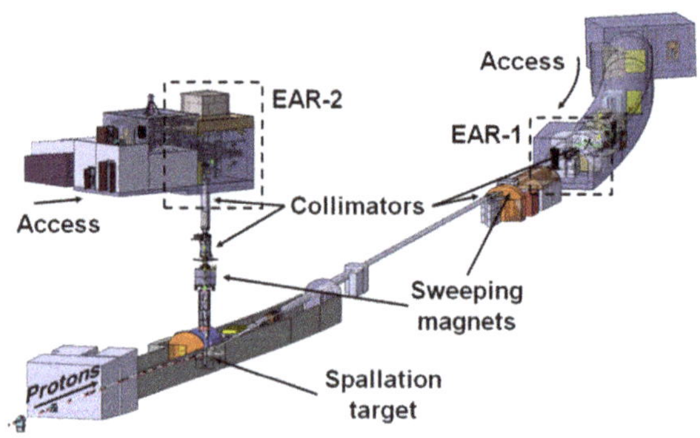

Figure 5.5. Layout of the nToF facility at CERN. The 20 GeV proton beam is incident from the bottom left on the spallation target. Neutrons are produced in all directions but a very small fraction are transported down the 185 m beamline to EAR1 and vertically via the 19 m beamline to EAR2. (Reproduced from [11]. CC-BY-4.0.)

The original nToF facility comprises a 185 m long neutron beamline reaching experimental area 1 (EAR1). It was later found to be desirable to have a shorter, vertical, 19 m, beamline to experimental area 2 (EAR2) where much higher neutron fluxes could be achieved. Figure 5.5 provides a general overview of the nToF infrastructure at CERN, while figure 5.6 shows the flux of neutrons (and gamma-rays) at EAR1 and EAR2 as a function of energy.

In order to understand the motivation for the design of such a time-of-flight facility and its achievable energy resolution, let us look in more detail at the underlying physics. First, we make the assumption that the high energy proton beam strikes the spallation target in a well-defined pulse; at nToF this condition is satisfied

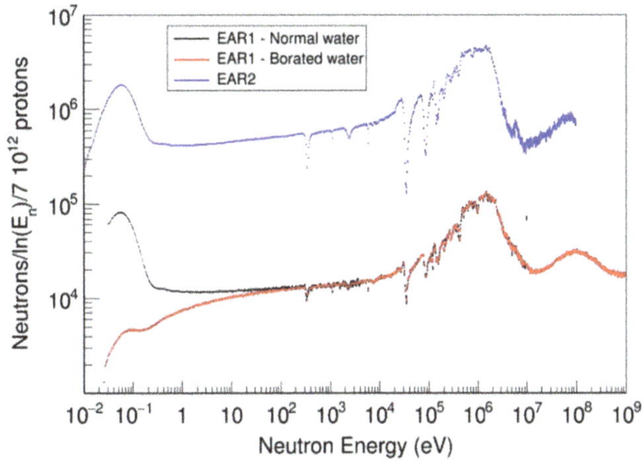

Figure 5.6. Energy distribution of neutrons at the two nToF experimental areas EAR1 and EAR2. The flux at EAR2 is much higher because it is ten times closer to the neutron production target. The flux is given as neutrons per proton beam pulse where each pulse contain $\sim 7 \times 10^{12}$ protons. The flux varies depending on whether water or borated water is used as the moderator. (Reproduced from [11]. CC-BY-4.0.)

as the beam pulse has a time duration of 7 ns (r.m.s.). The neutrons then fly in all directions but we can consider collimating them to produce a directional beam. The neutrons will have a wide range of energies (see figure 5.6). The kinetic energy of a neutron, E_n, travelling with velocity v is given by

$$E_n = E - mc^2 = \sqrt{c^2p^2 + m^2c^4} - mc^2 = mc^2(\gamma - 1), \tag{5.1}$$

where c is the speed of light and $\gamma = \frac{1}{\sqrt{1 - v^2/c^2}}$. For neutrons with energies below a few MeV, equation (5.1) can be rewritten in a classical form

$$E_n = \frac{1}{2}mv^2 \propto \frac{L^2}{t^2}, \tag{5.2}$$

where it is more evident how neutron kinetic energies are related to flight distance, L, and time-of-flight, t. Over a fixed distance, the neutrons will separate out by time-of-flight. The achievable energy resolution of a neutron time-of-flight system can be derived from equation (5.2) as follows:

$$\frac{\Delta E_n}{E_n} = 2 \times \sqrt{\left(\frac{\delta t}{t}\right)^2 + \left(\frac{\Delta L}{L}\right)^2}. \tag{5.3}$$

This demonstrates that the best neutron energy resolution will be achieved by producing the neutrons in as short a pulse as possible and in as compact a space as possible. Moreover, the flight distance and, hence, flight time should be as long as possible. This comes at a price, however, since simple geometric considerations would show that the effective solid angle of the neutron beam will be a very tiny fraction of the total neutron flux coming out of the spallation target; the vast

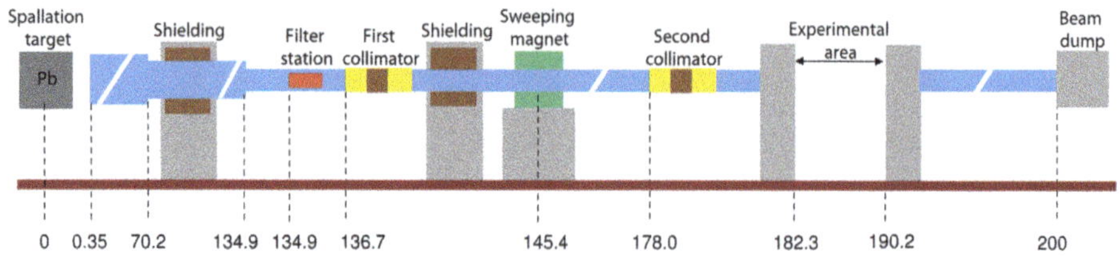

Figure 5.7. Layout of the nToF neutron beamline from the spallation target to the beam dump. Distances along the beamline are given in metres. (Reproduced from [10]. CC-BY 2.0.)

majority of neutrons produced are unavoidably dumped into the ground. Along the long flight path to EAR1, the neutron beam is collimated and there is a sweeping magnet to remove charged particles (see figure 5.7); exotic charged particles such as pions are produced when the high energy protons hit the spallation target. At EAR1, there is an experimental area which can accommodate various different types of detection apparatus surrounding a point where solid target materials can be mounted of the isotope of interest for a given experimental study. Permission exists to handle samples of radioactive material such as minor actinides.

Having given a general outline of the nToF facility, we now turn to some specific case studies that give a flavour of the types of nuclear data studies possible at a facility such as nToF.

5.2.2 Case study 1: ^{238}U(n,γ)

As discussed in section 5.1.1, neutron capture on the majority uranium isotope, ^{238}U, is highly important in terms of controlling the build-up of plutonium and the minor actinide inventory within a fission reactor (see figure 5.1). It is therefore desirable to understand the cross section for this reaction over a wide energy range as precisely and accurately as possible. Achieving this is extremely challenging, requiring not only long running times in order to reduce statistical uncertainty but also mandating strong constraints on potential sources of systematic error. Here, we present a case study related to the ^{238}U(n,γ) capture reaction carried out at nToF in the early 2010s [15, 16]. The motivations for the work at that time were critical uncertainties in the ^{238}U(n,γ) reaction cross section highlighted in the NEA high priority request list. Three energy regions were highlighted: 22.6 $\rightarrow$ 454 eV where the uncertainty at that time was 2% while the uncertainty asked for on the NEA request list was 1%, 2.03 $\rightarrow$ 9.12 keV (3% uncertainty versus requested 1%), and 9.12 $\rightarrow$ 24.8 keV (9% uncertainty versus requested 3%) [16]. The lowest energy region is within the so-called *resolved resonance region* (RRR) where individual resonances successively dominate the capture cross section; such resonances correspond to states in the compound nucleus (^{238}U + n $\rightarrow$ ^{239}U*). The higher energy regions requested begin to enter the *unresolved resonance region* where the level density in the compound nucleus is so high that resonances start to overlap as their intrinsic widths are comparable to the resonance spacing. Ultimately, all resonance structure is lost as resonances fully overlap.

A study was carried out at the nToF facility [15, 16] using a thin sample of very pure (99.999%) ^{238}U placed at the EAR1 target position so as to intercept 97% of the incoming neutron beam. The key to such a capture measurement is the detection of gamma-rays with high efficiency. The requirements in this respect differ from nuclear structure studies using gamma-ray spectroscopy since the precise details of the gamma-ray decay from individual resonant states is not important. However, coarse sensitivity to the total gamma-ray energy deposited and to the gamma-ray multiplicity may be a means of reducing backgrounds.

There are two significant challenges to gamma-ray detection in play at nToF. First, there is the so-called 'gamma flash' which precedes the arrival of the neutrons and is largely attributable to the decay of neutral pions produced in the primary beam–target interaction. Such pions are forward boosted so this effect is most important for EAR1 which is downstream of the primary production target. The second challenge is that of detecting gamma-rays in an environment where there is a high number of scattered neutrons given that most common classes of gamma-ray detectors cannot readily discriminate between gamma-ray and neutron interactions. Such neutrons can be removed as far as possible by a neutron moderator/absorber shell between the secondary target and the gamma-ray detectors. The most suitable absorber materials are those rich in ^{6}Li or ^{10}B since these isotopes have a large cross section for absorbing thermal neutrons and breaking up into a pair of light charge particles. Clearly, an absorber which captured thermal neutrons and then itself emitted gamma-rays would be a very poor choice in the vicinity of a gamma-ray detector.

Two principal approaches are taken to measurement of capture gamma-rays for (n,γ) measurements at nToF. The first is to use a pair of low efficiency, low energy resolution C_6D_6 detectors [17]. While such detectors have a poor performance overall for gamma-ray detection compared with more common types of detector, they have the distinct advantage in this application of having a very low sensitivity to neutrons. The second approach for (n,γ) measurements is to use a total absorption calorimeter (TAC [18]) to detect the entire γ-ray cascade emitted after neutron capture with as high an efficiency as possible using an array with as complete a solid angle coverage as possible. Such a detection approach has strong similarities in its philosophy to the total absorption gamma spectrometers (TAGS) described in section 3.3. The nToF TAC comprises 40 BaF_2 detectors of varying hexagonal and pentagonal shape to form an interlocking shell that covers 95% of the 4π solid angle (see figure 5.8). BaF_2 is an inorganic scintillator (described in section 5.1.2 of [14]) which, unusually, has two significant components to its scintillation light output: a slow component and a fast component with differing wavelength ranges. The fast timing component with sub-ns decay time improves the timing resolution of the system, while the energy resolution of BaF_2 ($\sim$13% at 1 MeV) is sufficient to make crude selections on reaction Q-value. Pulse-shape data from the gamma-ray detectors are captured with a digital data acquisition system [19]; the digitized waveforms are subsequently processed in the offline analysis phase.

The total energy expected to be emitted in a capture reaction involving some specific resonance in the resolved resonance region will be slightly larger than the

Figure 5.8. One hemisphere of the BaF_2 TAC array used in conjunction with nToF at CERN. Immediately surrounding the beampipe at the centre of the image is a neutron absorber shell. (Reproduced with permission from [18]. Copyright 2009 Elsevier.)

capture reaction Q-value; in the case of the ^{238}U(n,γ) reaction, the Q-value is 4806.38 (17) keV. In general, this energy will be emitted as a cascade of two or more gamma-rays from the capture resonance through an intermediate state (or states) and then to the ground state[4]. These specific characteristics of the decay provide a means of strongly suppressing background by requiring that valid events require more than one BaF_2 detector to have fired in prompt coincidence, as well as the total energy deposited to be between 2.5 and 5.75 MeV. This suppresses the strong background from neutron capture on hydrogen at around 2.2 MeV as well as a significant fraction of scattered neutron events. The effectiveness of this condition in suppressing background in the raw data is shown in figure 5.9.

There are many additional sources of background which need to be carefully handled. One relates to pile-up and dead time in the gamma-ray detectors due to the high counting rate (neutron flux) and the activity of the ^{238}U sample. Such generic, detector-related issues are explained in more detail in chapter 10 of [14]. Other background contributions can be assessed through measurements without the beam to understand long-lived gamma-ray backgrounds, and with the beam on and the sample removed (see figure 5.10). More detailed discussion of these issues is found in [15, 16].

Fundamentally, the parameter to be extracted from the measurement is the energy-dependent reaction yield $Y(E_n)$ given by

$$Y(E_n) = \frac{C(E_n) - B(E_n)}{\varepsilon \cdot F \cdot \phi_n(E_n)}, \quad (5.4)$$

[4] To find details of the known gamma-ray cascades following ^{238}U(n,γ) capture reactions, the reader is invited to search for ^{239}U on the ENSDF database.

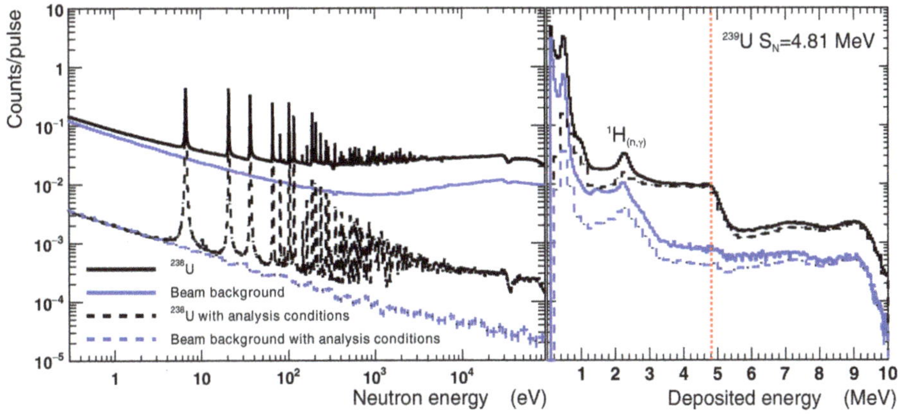

Figure 5.9. Left: Raw counts per histogram bin with (black) and without (blue) the ^{238}U target in place. The dashed lines are where the analysis conditions of the total energy and multiplicity in the BaF_2 array are applied (see the text for details). Right: Total energy deposited in the BaF_2 array for events in the neutron energy range 1–10 keV. (Reproduced from [16]. CC-BY 4.0.)

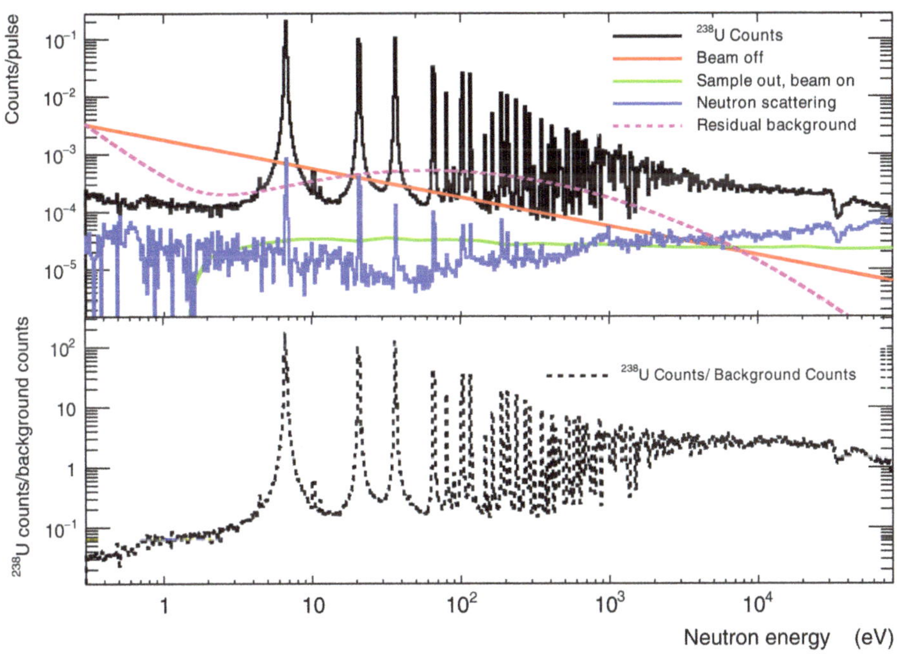

Figure 5.10. Background subtracted ^{238}U counts are compared to the individual and total background contributions. Neutron scattering also shows a resonant structure since the resonances in that reaction process largely correspond to those seen in neutron capture. (Reproduced from [16]. CC-BY 4.0.)

where $C(E_n)$ is the total counts, $B(E_n)$ is the background counts, ε is the efficiency of the detection system for measuring capture cascades, F is the fraction of the beam intercepted by the ^{238}U target and $\phi_n(E_n)$ is the number of incident neutrons in the given energy bin [15]. In practice, the experimental data are compared with the yield predicted by the *R*-matrix code, SAMMY, for the known resonances in order to achieve a normalization. The principles behind such an *R*-matrix approach were outlined earlier in section 1.2.9.

Exercises

5.5. Some scintillator detectors such as $LaBr_3$ offer high energy and timing resolution. Consider what benefits in sensitivity might have been achieved if $LaBr_3$ detectors were used instead of BaF_2 detectors in the $^{238}U(n,\gamma)$ study. (Looking up the parameters of such detectors online or accessing the relevant literature may help in answering this question.)

5.2.3 Case study 2: neutron-induced fission—$^{235}U(n,f)$

The other main area in which nToF contributes to improvements to knowledge of nuclear data is in the measurement of neutron-induced fission (n,f) reaction cross sections. Here, the challenges in experimental technique and data analysis are somewhat different from those discussed above for neutron capture reactions. We therefore present a case study focussing on neutron-induced fission reactors taking as an example a study of the $^{235}U(n,f)$ carried out at the nToF facility by Amaducci e*t al*; the reader is directed to the relevant publication for fuller details [13].

The $^{235}U(n,f)$ reaction underpins the operation of conventional fission reactors and so, it is perhaps, unsurprising that its cross section is extremely well-studied. It is considered a neutron-induced reaction standard both at thermal energies and in the energy range 150 keV to 200 MeV (see table 5.2). Prior to the publication of the work described here [13], the cross section was less well standardized in the energy range between thermal energies and 150 keV. Moreover, some of the published data in the energy range 10–30 keV pointed to inconsistencies at the level of a few percent with the evaluated data. This level of disagreement might seem small but given the high importance of the $^{235}U(n,f)$ reaction and its cross section being used to

Table 5.2. Reactions and the corresponding energy range in which they correspond to neutron cross section standards as reported in the 2017 evaluation [12]. 'MACS' stands for Maxwellian averaged cross section.

Reaction	Incident neutron energy range
H(n,n)	1 keV to 20 MeV
$^{3}He(n,p)$	0.0253 eV to 50 keV
$^{6}Li(n,t)$	0.0253 eV to 1 MeV
$^{10}B(n,\alpha)$	0.0253 eV to 1 MeV
$^{10}B(n,\alpha_1\gamma)$	0.0253 eV to 1 MeV
C(n,n)	10 eV to 1.8 MeV
$Au(n,\gamma)$	0.0253 eV, 0.2 to 2.5 MeV, 30 keV MACS
$^{235}U(n,f)$	0.0253 eV, 7.8–11 eV, 0.15–200 MeV
$^{238}U(n,f)$	2–200 MeV

characterize various facilities including nToF, removing these uncertainties is important.

In contrast to charged particles such as alpha particles or heavy-ion beams, neutrons are highly penetrating. This allows for rather different experimental arrangements than what might be typically found in a nuclear spectroscopy measurement such as those described in chapter 6 of [14] or in the production of medical isotopes with charged-particle beams described in chapter 6 of the present work. Given that neutrons may readily propagate through materials with limited interaction, it is favourable to deploy a stack of target samples and use some of them to standardize the measurement of the sample whose cross section is either unknown or which wants to measured with higher higher precision and accuracy. Here, the standardization was performed with respect to two reference reactions ^{6}Li(n,t) and ^{10}B(n,α) which are well characterized through the energy range of interest (see figure 5.11). As noted in section 9.2.2 of [14], these reactions are commonly used as the basis of thermal neutron detectors in replacement of relatively expensive ^{3}He. As an aside, it is worth mentioning that ^{6}Li and ^{10}B are very unusual isotopes in that they have an odd number of both protons and neutrons. Even–even nuclei experience a gain in binding energy from like-nucleon pairing while isotopes with odd numbers of protons and neutrons are a much less stable configuration (see section 1.1.1). In fact, ^{6}Li and ^{10}B are two of only a handful of stable odd–odd isotopes which exist in nature.

In terms of the experimental arrangement for this study, a series of target samples of ^{6}Li, ^{10}B and ^{235}U deposited on aluminium foils were mounted in the neutron beam (see figure 5.12). Between each successive target foil are two 200 μm thick silicon detectors (see chapter 8 of [14] for a description of silicon detectors and how they work). These detectors record charged particles emitted following neutron-induced reactions in the respective target foils. In the case of ^{235}U, these will be fission fragments and in the case of ^{6}Li and ^{10}B, these will be light ions such as tritons and alpha particles. The silicon detectors are doubled in order to detect forward-going (i.e. neutron beam direction) and backward-going charged particles in order to reduce uncertainties related to the angular distribution of emitted charged particles, particularly for the ^{6}Li(n,t) and ^{10}B(n,α) reactions. An experimental challenge is the differing energy range of emitted fission fragments (up to ~100 MeV) and break-up reaction products from reactions on ^{6}Li and ^{10}B which are more in the range of a few MeV. This is handled by instrumenting the silicon detectors with logarithmic preamplifiers which are linear up to 10 MeV. Good separation of fission fragments from alpha particles (the principal alpha decay of ^{235}U has an energy of 4.679 MeV) is achievable using this system as shown in figure 5.13.

The technique used to extract the ^{235}U(n,f) cross section is the so-called 'ratio method'. This expresses the cross section in terms of the cross section of the reference reactions on e.g. ^{10}B [13]:

$$\sigma_{^{235}\mathrm{U}} = \frac{C_{^{235}\mathrm{U}} f_{\mathrm{ref}}\, \rho_{\mathrm{ref}}\, \varepsilon_{\mathrm{ref}}}{C_{\mathrm{ref}} f_{^{235}\mathrm{U}}\, \rho_{^{235}\mathrm{U}}\, \varepsilon_{^{235}\mathrm{U}}} \sigma_{\mathrm{ref}}, \tag{5.5}$$

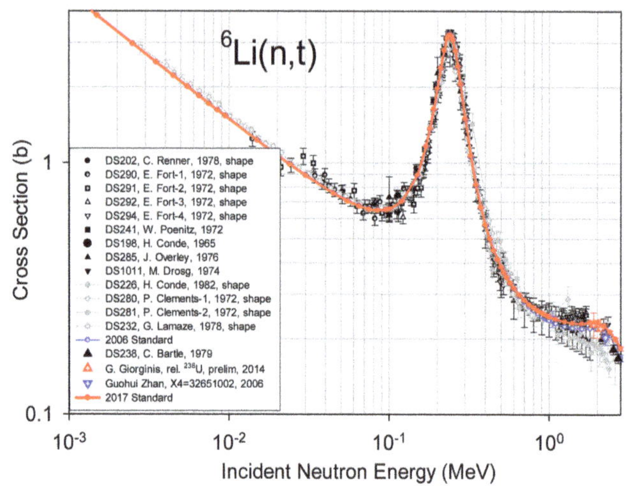

(a) Incident neutron energies from 1 keV to 3 MeV.

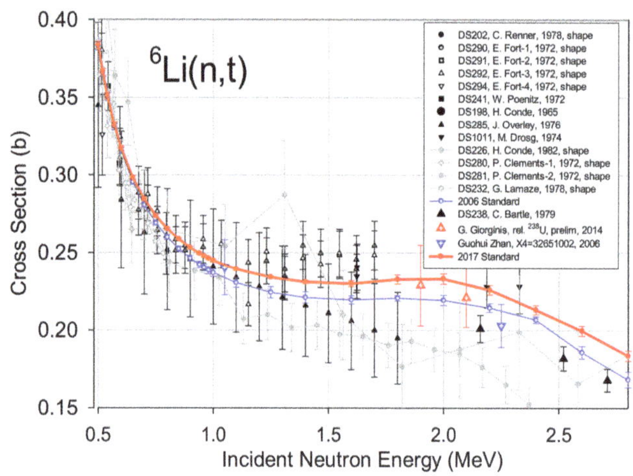

(b) Incident neutron energies from 0.5 MeV to 2.8 MeV.

Figure 5.11. Comparison of 2017 and 2006 neutron standard cross sections for the ^{6}Li(n,t) reaction. The data points represent different experimental measurements (see [12] for details). (Reproduced with permission from [12]. Copyright 2018. Published by Elsevier Inc. CC-BY-NC-ND 4.0.)

where $\sigma_{^{235}\mathrm{U}}$ and σ_{ref} are the cross sections for ^{235}U(n,f) and the reference reaction, respectively, $C_{^{235}\mathrm{U}}$ and C_{ref} are the respective number of counts recorded, $\rho_{^{235}\mathrm{U}}$ and ρ_{ref} are the respective areal densities of the targets, $\varepsilon_{^{235}\mathrm{U}}$ and $\varepsilon_{\mathrm{ref}}$ are the respective detection efficiencies for the outgoing reaction products, and, finally, $f_{^{235}\mathrm{U}}$ and f_{ref} are the respective fractions of the neutron beam intercepting the targets. Naturally, all these parameters are neutron energy-dependent with the exception of the target areal densities. Some of these parameters such as the detection efficiencies require Monte Carlo modelling. Parameters which needs particular attention are $f_{^{235}\mathrm{U}}$ and

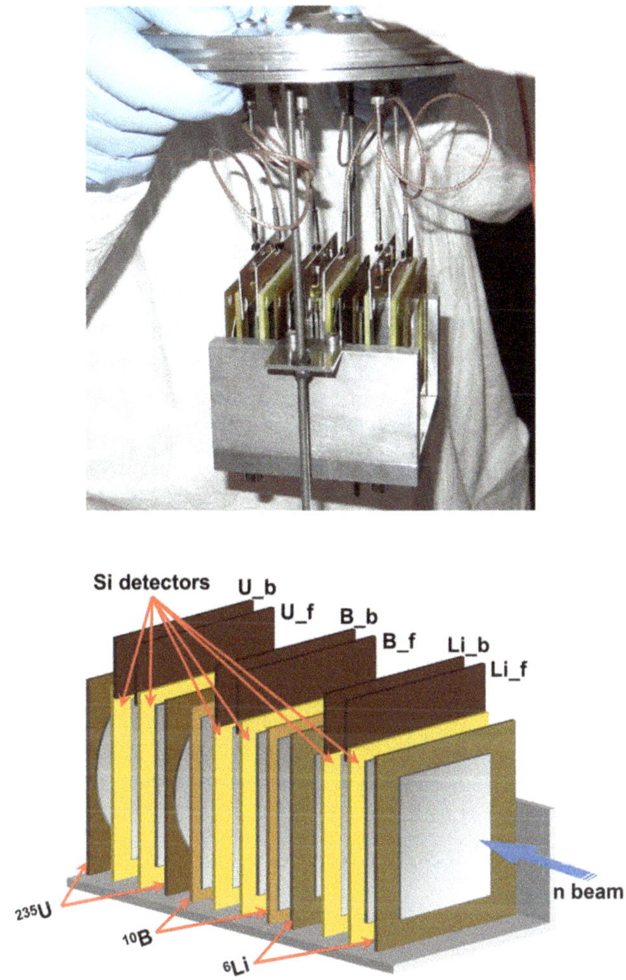

Figure 5.12. Target and detection system used for measurement of the $^{235}U(n,f)$ reaction. A stack of thin aluminium foils with thing deposited samples of ^{6}Li, ^{10}B and ^{235}U on them, interspersed with 200 μm thick silicon detectors. (Reproduced from [13]. CC-BY-4.0.)

f_{ref}—the fraction of the neutron beam intercepting the respective targets. While attenuation of the neutron beam by the targets initially intercepted along the trajectory is small, it is certainly not negligible. This is one of the more important aspects to handle in order to constrain systematic error.

In summary, the case study reported here used a clever technique to standardize one cross section with respect to others. This approach has wide utility in nuclear data studies. In this specific case, it allowed the $^{235}U(n,f)$ cross section to be obtained with high precision in the energy range 25 meV to 170 keV with a 1.5% systematic uncertainty (see figure 5.14).

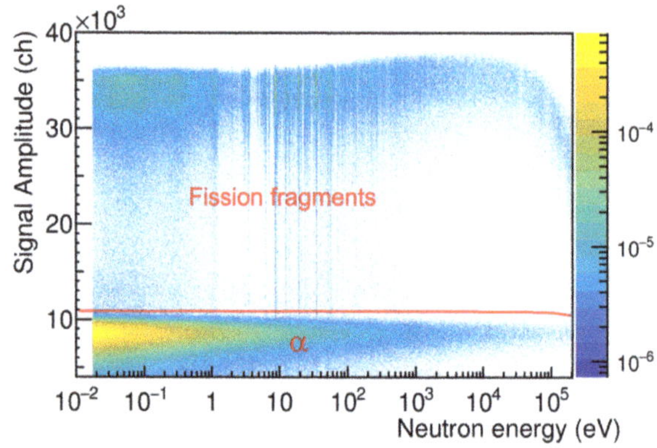

Figure 5.13. Separation of fission fragments from natural alpha decay of ^{235}U. A plot of silicon detector amplitude as a function of neutron energy. See the text for discussion of the employment of logarithmic preamplifiers. (Reproduced from [13]. CC-BY-4.0.)

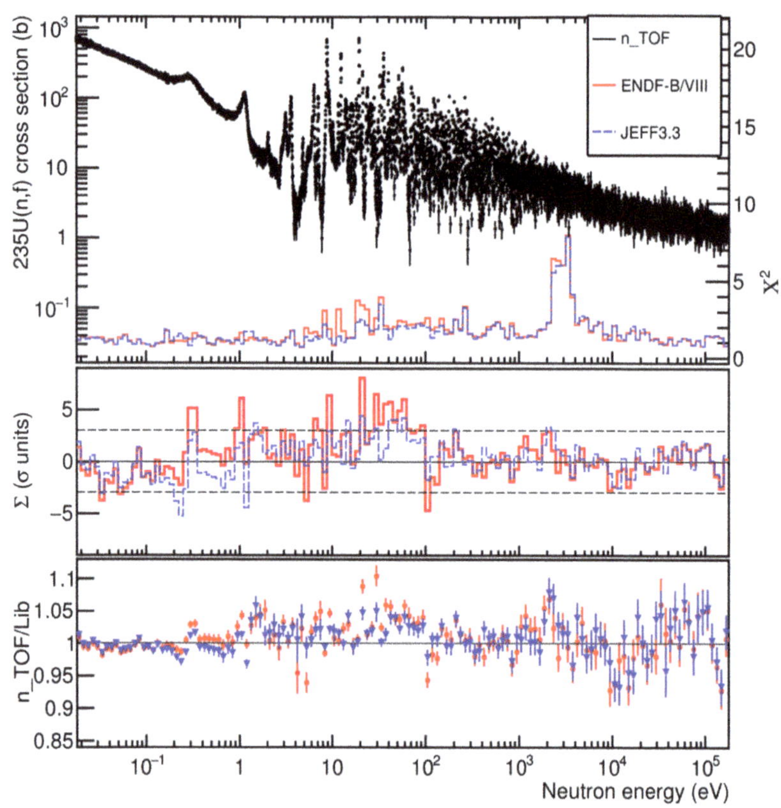

Figure 5.14. Comparison of the nToF measurement for the $^{235}U(n,f)$ reaction with the tabulated values found in the ENDF-B/VIII and JEFF3.3 databases. (Reproduced from [13]. CC-BY-4.0.)

Exercises

5.6. Consider extending the $^{235}U(n,f)$ methodology described here to studies of the neutron-induced fission of some of the minor actinides in figure 5.2. What limitations might you expect if samples are harder to access and the specific activity of samples is higher? How could the experimental arrangement be modified to take account of these changes?

5.3 Nuclear waste storage and disposal

We now turn to nuclear waste management and disposal—specifically the spent fuel from a fission reactor. This is one of the aspects of nuclear energy that introduces significant complication and expense. While, most of the relevant science in this regard is more properly nuclear engineering and nuclear chemistry, there are some areas where *nuclear data* are important. Here, we give two brief examples: one concerning neutron production within spent fuel and the second concerning nuclear transmutation.

5.3.1 Neutron generation within spent fuel

We noted in section 3.4 how delayed neutrons originating from β-delayed neutron emission of fission products are important in reactor control. Such neutrons are emitted over a timescale of minutes. We might therefore expect that on longer time scales when spent fuel has been removed from a reactor that there would be no further processes leading to neutron emission. Neutron emission is, of course, something that should be monitored because it could restart neutron-induced fission within the material. In practice, there are at least two mechanisms that can produce neutrons in spent fuel. The first would be spontaneous fission of minor actinides in the material. The second is perhaps more surprising since it involves nuclear reactions in the material, namely, (α,n) reactions driven by alpha particles emitted from minor actinides in the spent fuel. The range of such alpha particles is short, typically of order 10 microns in solids. Given this very short range, interactions are only feasible with other elements bound up within the fuel itself. Since nuclear fuels are commonly in the form of uranium oxide, the most relevant (α,n) reactions therefore typically occur on the neutron-rich isotopes of oxygen, namely, ^{17}O, ^{18}O. The reason that these isotopes are significant in neutron production and not the majority isotope, ^{16}O is down to reaction Q-value. In fact, the Q-value for the $^{17}O(\alpha,n_0)^{20}Ne$ reaction[5] is positive (+0.5881 MeV) meaning that there is no threshold for this reaction although due to the Coulomb barrier, the cross section at low energies is very small (see figure 5.15).

As shown in figure 5.2, the minor actinides typically have long half lives of tens to thousands of years. This means that the source of alpha particles is semi-continuous on a human timescale. In order to evaluate the yield of neutrons one has to consider

[5] The n_0 here means that the reaction proceeds to the ground state of ^{20}Ne.

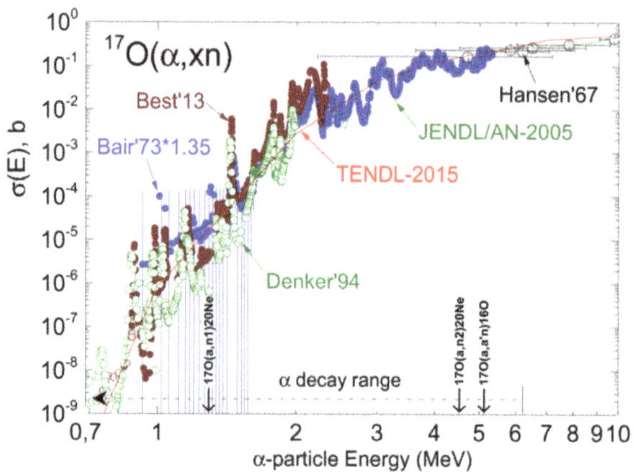

Figure 5.15. Compilation of cross section data for the ^{17}O(α,n) reaction from various earlier experimental measurements. The inconsistency of some of these data is obvious. The ^{17}O(α,n$_0$)^{20}Ne reaction has no threshold and is open at all energies; the location of thresholds for some other potential reactions involving ^{17}O is indicated. (Reproduced with permission from [20]. Copyright 2017 Elsevier.)

the relevant cross sections integrated down from the initial alpha particle energy to zero. The initial alpha particle energy will differ for each alpha-decaying isotope since the more exotic minor actinides such as ^{252}Cf emit higher energy alpha particles (see figure 5.16). The yield of neutrons for a single incident alpha particle with energy E_α is then given by

$$Y_n(E_\alpha) = N_T \int_0^{E_\alpha} \sigma_n(E)|S(E)|^{-1}\mathrm{d}E, \tag{5.6}$$

where N_T is the atomic density of the target nucleus, $\sigma_n(E)$ is the energy-dependent cross section of the target nucleus and $S(E)$ is the linear stopping power of the material (see discussion of stopping powers in appendix A).

5.3.2 Nuclear transmutation

Some degree of separation and reprocessing of fuel is possible; indeed some minor actinides retrieved in this manner such as ^{241}Am have found application in smoke detectors and more recently as 'space batteries', i.e. thermoelectric generators on space missions [21]. Ultimately, much of the waste will still require storage ideally in a geologically safe underground repository. Clearly, there are a whole range of engineering and safety challenges entailed in this, let alone the expense.

People have long been interested in whether one can find a technological solution to the problem of long-lived nuclear waste[6]. One proposed solution is *nuclear transmutation* where one employs nuclear reactions to transform the very long-lived

[6] In some sense, carbon capture technology forms an interesting parallel as to how one remediates environmentally damaging waste products.

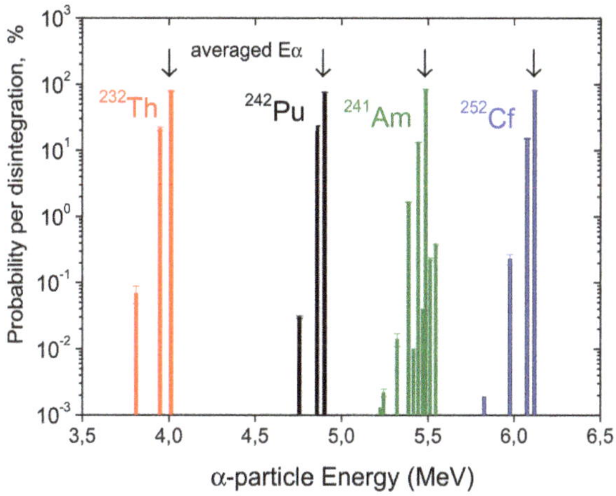

Figure 5.16. Energies of the characteristic alpha emissions of ^{232}Th, ^{242}Pu, ^{241}Am and ^{252}Cf. The average alpha emission energy is indicated. (Reproduced with permission from [20]. Copyright 2017 Elsevier.)

isotopes into shorter-lived ones. Many different schemes have been proposed for how to drive such conversion using, for example, a fast reactor [22, 23], accelerator driven fast reactors [24] or intense gamma-ray photon beams [25]. Essentially, one needs to drive nuclear reactions of the type (n,γ), (n,2n), (n,p), (γ,n) or (γ,p) that are going to transform challenging long-lived isotopes into neighbouring isotopes which decay more rapidly and through a different decay pathway. One can imagine that a complicated simulation will be required to understand how best to optimize such a process since it might work well for one isotope but poorly for others. In addition, the nuclear data required, particularly cross section data, may not be known with sufficient precision. Obtaining new data where the targets for such experiments are themselves radioactive is not straightforward. Lastly, such transmutation projects represent formidable challenges in engineering and energy input. Even in optimistic scenarios, it may take decades to effectively reduce the inventory of long-lived radioisotopes.

Exercises

5.7. Consider how to burn up one (or more) of the long-lived fission products shown in table 5.1. What would be the best way to go about that, i.e. choice of reactions? What other things might constrain the feasibility? Here we start to enter into a true research project. Some useful references as starting points are given in the text, i.e. [22–25].

References

[1] Salvatores M and Palmiotti G 2011 Radioactive waste partitioning and transmutation within advanced fuel cycles: achievements and challenges *Prog. Part. Nucl. Phys.* **66** 144–66

[2] Sublet J-C, Eastwood J W, Morgan J G, Gilbert M R, Fleming M and Fleming W 2017 FISPACT-II: an advanced simulation system for activation, transmutation and material modelling *Nucl. Data Sheets* **139** 77–137

[3] Abram T and Ion S 2008 Generation-IV nuclear power: a review of the state of the science *Energy Policy* **36** 4323–30

[4] Krall L M, Macfarlane A M and Ewing R C 2022 Nuclear waste from small modular reactors *Proc. Natl Acad. Sci.* **119** e2111833119

[5] Humphrey U E and Khandaker M U 2018 Viability of thorium-based nuclear fuel cycle for the next generation nuclear reactor: issues and prospects *Renew. Sustain. Energy Rev.* **97** 259–75

[6] van Rooijnen W F G, Shimazu Y and Yamano N 2015 Criticality uncertainty dependence on nuclear data library in fast molten salt reactors *Energy Proc.* **71** 3–13

[7] Zhu W, Wang Z, Zhu X, Shu N and Fan T 2017 A new evalutation of fission product yields for the neutron induced fission of U-233 and Th-232 *Fusion Eng. Des.* **125** 608–14

[8] Worth R N *et al* 2021 The distribution and selective decontamination of carbon-14 from nuclear graphite *J. Nucl. Mater.* **556** 153167

[9] Otuka N *et al* 2014 Towards a more complete and accurate experimental nuclear reaction data library (EXFOR): international collaboration between nuclear reaction data centres (NRDC) *Nucl. Data Sheets* **120** 272–6

[10] Guerrero C *et al* 2013 Performance of the neutron time-of-flight facility nTOF at CERN *Eur. Phys. J.* A **49** 27

[11] Colonna N *et al* 2020 The fission experimental programme at the CERN nTOF facility: status and perspectives *Eur. Phys. J.* A **56** 48

[12] Carlson A D *et al* 2018 Evaluation of the neutron data standards *Nucl. Data Sheets* **148** 143–88

[13] Amaducci S *et al* 2019 Measurement of the ^{235}U(n,f) cross section relative to the ^{6}Li(n,t) and ^{10}B(n,α) standards from thermal to 170 keV neutron energy range of nTOF *Eur. Phys. J.* A **55** 120

[14] Jenkins D 2020 *Radiation Detection for Nuclear Physics: Methods and Industrial Applications* (IOP Publishing)

[15] Wright T 2014 High accuracy measurement of the ^{238}U(n,γ) cross section at the CERN nTOF facility *PhD Thesis* University of Manchester

[16] Wright T *et al* 2017 Measurement of the ^{238}U(n,γ) cross section up to 80 keV with the Total Absorption Calorimeter at the CERN nTOF *Phys. Rev.* C **96** 064601

[17] Plag R, Heil M, Käppeler F, Pavlopoulos P, Reifarth R and Wisshak K 2003 An optimized C_6D_6 detector for studies of resonance-dominated (n,γ) cross-sections *Nucl. Instrum. Methods Phys. Res.* A **496** 425–36

[18] Guerrero C *et al* 2009 The nTOF total absorption calorimeter for neutron capture measurements at CERN *Nucl. Instrum. Methods Phys. Res.* A **608** 424–33

[19] Abbondanno A *et al* 2005 The data acquisition system of the neutron time-of-flight facility nTOF at CERN *Nucl. Instrum. Meth.* A **538** 695–702

[20] Simakov S P and van den Berg Q Y 2017 Updated of the α–n yields for reactor fuel materials in the interests of nuclear safeguards *Nucl. Data Sheets* **139** 190–203

[21] Dustin J S and Borrelli R A 2021 Modeling of Am-241 as an alternative fuel source for radioisotope thermoelectric generator *Nucl. Eng. Des.* **385** 111495

[22] Chiba S, Wakabayashi T, Tachi Y, Takaki N, Terashima A, Okumura S and Yoshida T 2017 Method to reduce long-lived fission products by nuclear transmutations with fast spectrum reactors *Sci. Rep.* **7** 13961
[23] Sun X Y, Han L H, Li X X, Hu B L, Luo W and Liu L 2023 Transmutation of MAs and LLFPs with a lead-cooled fast reactor *Sci. Rep.* **13** 1693
[24] Abderrahim H A, De Bruyn D, Van den Eynde G and Michiels S 2014 Transmutation of high-level nuclear waste by accelerator driven systems *WIREs Energy Environ.* **3** 60–9
[25] Sun X Y, Luo W, Lan H Y, Song Y M, Gao Q Y, Zhu Z C, Chen J G and Cai X Z 2022 Transmutation of long-lived fission products in an advanced nuclear energy system *Sci. Rep.* **12** 2240

Chapter 6

Nuclear data for the production of medical isotopes

Mamad Eslami

As our population ages, access to nuclear medicine is vital for the diagnosis and treatment of cancer and other diseases. Safe employment of nuclear medicine and the production of relevant medical isotopes requires precise and accurate nuclear data. Here, we review the connections between nuclear data and nuclear medicine with the aid of some topical case studies. We begin with a review of the main modalities used to visualize physiological processes and deliver targeted treatment, highlighting the physical principles behind each technique and the types of radiation involved. We then turn to the experimental methods used to measure nuclear reaction and decay data, and explain how these data support the safe and reliable production of medical radioisotopes. Two production case studies, ^{67}Cu and ^{18}F, show how measured cross sections underpin routine supply. Two further examples, involving the decay data of ^{152}Tb and ^{225}Ac, reveal the challenges of modelling complex decay schemes and cascade decay chains, and the need for accurate half-lives and branching ratios. Finally, we examine how international nuclear data centres, such as the IAEA Nuclear Data Services (NDS) and the Evaluated Nuclear Data File (ENDF), keep this information current and accessible to the medical community.

Concepts: Modalities in nuclear medicine, PET, SPECT, therapy, theranostics, experimental methods, accelerators e.g. cyclotrons and linacs, gamma-ray spectroscopy and cross section evaluation, decay data, data compilation and evaluation

6.1 Modalities in nuclear medicine

Nuclear medicine is a field of medical imaging and therapy in which radioactive isotopes are introduced into the body to probe physiological function or to selectively destroy diseased tissue. Unlike external beam treatments such as proton or ion therapy—where charged particles are delivered from outside the body to

doi:10.1088/978-0-7503-5102-7ch6 © IOP Publishing Ltd 2025. All rights, including for text and data mining (TDM), artificial intelligence (AI) training, and similar technologies, are reserved.

irradiate tumours—nuclear medicine involves the internal administration of *radiopharmaceuticals*—biologically active compounds labelled with radionuclides that emit radiation from within. These radiotracers are designed to target specific cellular or molecular processes, enabling both diagnosis and treatment at the functional level. A typical radiopharmaceutical consists of a radioactive isotope coordinated to a chelator, which is chemically linked to a biological targeting vector such as a peptide, antibody, or small molecule.

The techniques of nuclear medicine fall into two broad categories: *diagnostic* and *therapeutic*. In diagnostic applications, isotopes emit gamma-rays[1] that can be detected externally, allowing clinicians to image metabolism, receptor expression, or blood flow *in vivo*. Therapeutic applications instead exploit the energy deposited by β^- particles, α particles, or Auger electrons to induce lethal damage to targeted cells, often through DNA double-strand breaks. A growing approach involves using the same targeting compound for both imaging and therapy by labelling it with different radionuclides that emit either diagnostic or therapeutic radiation.

At the heart of all these applications lies the need for high-quality *nuclear data*. Reaction cross sections determine how efficiently radionuclides can be produced in accelerators or reactors. Decay half-lives must align with biological uptake and clearance times, while emission probabilities and energy spectra influence detector response, dosimetric accuracy and clinical safety margins. From production and radiolabelling to imaging, treatment planning, and regulatory approval, each stage in the use of radiopharmaceuticals depends on the availability and accuracy of these data. Here, we begin with an overview of the major clinical modalities in nuclear medicine, paying particular attention to the nuclear data required to support each.

6.1.1 PET

Positron emission tomography (PET) is a non-invasive imaging technique that visualizes the *in vivo* distribution of radiotracers labelled with positron-emitting radionuclides. Following β^+ decay, the emitted positron travels through tissue, losing kinetic energy gradually through ionization and excitation interactions with surrounding electrons. In soft tissue or water—often treated as radiological equivalents—the range of a positron varies from sub-millimetre to several millimetres, depending on its initial energy. Positrons annihilate with electrons, most commonly producing two gamma photons of 511 keV that are emitted in nearly opposite directions, thereby conserving both energy and linear momentum in the centre-of-mass frame. Such annihilation may occur directly while the positrons are slowing in the medium, or via the formation of a transient *positronium* atom—a bound state of the positron and an electron.

PET imaging relies on detection of annihilation photons in temporal coincidence by a ring of scintillation detectors surrounding the patient. Each event defines a line of response (LOR) along which the annihilation is assumed to have occurred.

[1] In some cases, such as PET isotopes, a positron is emitted that latterly annihilates into two (or more) gamma-rays.

A large number of such events are accumulated to reconstruct the spatial distribution of the radiotracer using tomographic algorithms. Spatial resolution in PET is influenced by several physical and instrumental factors, including positron range, non-collinearity of annihilation photons, Compton scattering and detector timing resolution. Time-of-flight (TOF) capabilities can improve localization along the LOR, and attenuation correction is typically performed using co-registered CT or MRI images. Emerging approaches are exploring how to exploit the quantum properties of the annihilation photons in PET imaging. In particular, the 511 keV photon pair produced from electron–positron annihilation is entangled in polarization, and several recent studies have demonstrated that this entanglement can persist even after one of the photons undergoes Compton scattering [1–3]. These results challenge the long-standing assumption that Compton scattering destroys the entanglement of annihilation photons, and suggest that even scattered photon pairs may retain diagnostic value. This opens the possibility of using events that are currently discarded in conventional PET—such as those involving intermediate scatter—to extract tissue-specific information based on residual polarization correlations, potentially enabling new modes of quantum-enhanced imaging. Prototype PET systems, such as J-PET, are already investigating how entangled-photon or positronium lifetime observables might be used to enhance diagnostic contrast [4, 5], although these technologies remain in the experimental stage.

Apart from detector and reconstruction innovations, the choice of radionuclide also has a strong impact on image resolution, largely due to the energy—and hence tissue range—of the emitted positron. Short-range positron emitters, such as ^{11}C, ^{18}F and ^{64}Cu, provide superior spatial resolution. In contrast, longer-range positron emitters like ^{13}N, ^{68}Ga and ^{124}I tend to result in greater blurring due to extended positron tracks. Among these, ^{18}F remains the most widely used isotope due to its favourable physical half-life and low mean positron energy.

6.1.2 SPECT

Single photon emission computed tomography (SPECT) is a widely used nuclear imaging technique that reconstructs three-dimensional distributions of gamma-emitting radiotracers based on the detection of individual photons emitted from within the patient's body. Unlike PET, which relies on coincident detection of annihilation photons, SPECT captures single gamma photons emitted directly from nuclear transitions, such as isomeric decay or gamma emission following β^- decay. This allows the use of a broader range of radionuclides with varying gamma energies and half-lives, matched to different clinical requirements.

The most commonly used SPECT radionuclide is ^{99m}Tc, a metastable nuclear isomer that decays via an isomeric transition to ^{99}Tc, emitting a 140.5 keV gamma-ray with high intensity [6]. Its physical and chemical properties—including a 6 hour half-life, favourable gamma energy for collimated detection and compatibility with a wide variety of chelators—make it the foundation of routine diagnostic nuclear medicine. Historically, ^{99m}Tc has been produced in generators as a decay product of the longer-lived ^{99}Mo. The latter is readily produced as a fission product; however, only a few reactors worldwide have the correct set-up and conditions to extract ^{99}Mo. At times, this has been seen as a major risk and has led to efforts to develop alternative production routes using accelerators [6]. Exploring such alternatives draws heavily on nuclear data and identifies requirements for new data.

Other SPECT isotopes include ^{123}I, ^{111}In and ^{201}Tl, as well as therapeutic isotopes such as ^{131}I and ^{67}Cu, which emit gamma photons that allow for imaging during treatment.

SPECT acquisition is performed using gamma cameras equipped with parallel-hole collimators that provide angular resolution through mechanical collimation. As the cameras rotate around the patient, projection data are collected from multiple angles and used to reconstruct the tracer distribution tomographically. In comparison to PET, SPECT generally has lower sensitivity and spatial resolution due to intrinsic losses from collimation and photon attenuation. However, it benefits from greater tracer versatility, lower system cost and long-established clinical protocols. Hybrid SPECT/CT systems are commonly used to provide anatomical localization and attenuation correction.

Recent advances in detector technology, particularly the use of cadmium zinc telluride (CZT) solid-state detectors, have improved energy resolution and sensitivity, expanding clinical applications. One such development is dual-isotope imaging, where tracers such as ^{99m}Tc and ^{123}I or ^{111}In are imaged simultaneously. If their gamma energies are sufficiently distinct, their distributions can be separated within the same acquisition. Accurate signal separation requires correction for Compton scatter and photopeak cross-talk, using multi-window techniques or Monte Carlo based methods [7].

Alternative SPECT geometries have also been developed to overcome the limitations of conventional collimation. These include coded-aperture systems, multi-pinhole and slit-slat collimators, all designed to enhance sensitivity and spatial resolution through different forms of geometric and computational optimization [8]. More radically, Compton cameras eliminate collimators entirely, reconstructing the photon origin along a cone by tracking sequential scatter and absorption events in layered detectors. These systems enable imaging across a broad energy range (100 keV–1 MeV) and show promise in applications requiring high sensitivity, fast acquisition, or imaging of high-energy emitters [9]. Although largely in the research phase, these innovations highlight the evolving capabilities of SPECT beyond its conventional designs, with potential impact on multi-tracer protocols, dynamic imaging and quantitative dosimetry.

6.1.3 Therapeutic isotopes

Radionuclide therapy is a form of internal radiation treatment in which radioactive isotopes are used to deliver cytotoxic doses of ionizing radiation to cancer cells. Unlike external beam therapy, the radiation is emitted from within the body, allowing more selective irradiation of the target. The clinical use of therapeutic isotopes depends on several physical and chemical factors, including the type and energy of radiation emitted, its penetration range in tissue, and the ability to stably bind the radionuclide to a biological carrier.

6.1.3.1 *Beta emitters*

Most therapeutic isotopes emit β^- particles that travel a few millimetres in tissue and cause damage primarily through ionization. One of the most established examples is ^{131}I, widely used to treat thyroid cancer and hyperthyroidism. It has a half-life of 8.0 days and emits β^- particles with a mean energy of 182 keV for therapy, along with 364.4 keV gamma photons with 81.5% intensity for imaging.

A widely used alternative is ^{177}Lu, particularly in peptide receptor radionuclide therapy (PRRT) for neuroendocrine tumours. With a half-life of 6.6 days, it emits β^- particles with a mean energy of 133.6 keV and a tissue range of approximately 1–2 mm—ideal for small lesions and micrometastases. Its decay also produces gamma photons at 113 keV (6.2% intensity) and 208 keV (10.4% intensity), which enable post-treatment SPECT imaging. However, the low photon intensity means conventional SPECT systems often require extended acquisition times to achieve sufficient image quality. As discussed earlier, recent advances in SPECT technology, including CZT detectors and improved collimation, have enhanced energy resolution and sensitivity, helping to overcome these limitations. This has enabled more efficient and quantitative imaging with ^{177}Lu, supporting its growing use in theranostic protocols that rely on image-based personalized dosimetry. CZT-based systems, in particular, have demonstrated the feasibility of whole-body post-treatment imaging with ^{177}Lu, achieving clinically acceptable acquisition times and quantification accuracy [10].

For larger or more diffuse tumours, ^{90}Y is often preferred due to its high β^- energy, which allows for deeper tissue penetration. However, as it does not emit gamma photons suitable for conventional imaging, post-therapy dosimetry relies on indirect detection methods. One approach is bremsstrahlung SPECT, which captures the broad-spectrum x-rays produced as β^- particles decelerate in tissue. While this allows visualization of distribution, its low spatial resolution and complex scatter background limit quantitative accuracy. An alternative is ^{90}Y PET, which exploits the rare internal pair transition—a process in which a virtual gamma transition between nuclear states above 1.022 MeV directly produces an electron–positron pair. Despite a low branching ratio of around 0.0034%, modern TOF-PET systems can produce clinically useful images with spatial resolution comparable to ^{18}F PET. Both methods, although limited by low photon statistics, have been successfully applied to assess microsphere distribution and estimate absorbed doses in patients treated with ^{90}Y [11, 12].

6.1.3.2 *Alpha therapy*

Targeted alpha therapy (TAT) is a highly localized form of radionuclide therapy that uses the physical characteristics of alpha particles to destroy cancer cells with extraordinary precision. The kinetic energy of emitted alpha particles generally falls between 4 and 9 MeV, reflecting the energy differences between nuclear states in alpha-decaying heavy nuclei. Due to their large mass and double positive charge, alpha particles interact strongly with matter and lose energy rapidly, travelling only 40→100 μm in tissue—approximately the diameter of a few cells. This short range, combined with high linear energy transfer (LET), allows them to deposit large amounts of energy over a small distance, producing dense ionization tracks.

Such ionization results in complex, often irreparable DNA damage, making alpha particles exceptionally effective at killing individual tumour cells.

TAT is particularly well suited for eliminating minimal residual disease, micrometastases, and circulating tumour cells—clinical scenarios where complete eradication of small, dispersed cancer populations is vital. Several alpha-emitting radionuclides have been developed for therapeutic use, with differing decay characteristics and chemistry:

- **^{225}Ac** ($t_{1/2} = 9.9$ d): Undergoes a cascade of four alpha decays, delivering high cumulative energy. However, each alpha emission imparts significant recoil to the daughter nucleus, often breaking chemical bonds and leading to redistribution of radioactive progeny such as ^{213}Bi.
- **^{211}At** ($t_{1/2} = 7.2$ h): A halogen that decays via alpha emission and electron capture. Its chemical similarity to iodine facilitates labelling of small organic molecules and peptides.
- **^{223}Ra** ($t_{1/2} = 11.4$ d): A calcium mimetic that naturally localizes to bone. ^{223}Ra dichloride, marketed as Xofigo, is currently the only alpha-emitting radiopharmaceutical approved for clinical use. It was authorized in 2013 by the United States Food and Drug Administration (FDA) and the European Medicines Agency (EMA) for treating metastatic castration-resistant prostate cancer (mCRPC) with bone involvement [13].

Many alpha emitters decay through complex chains involving both alpha and beta transitions. While these chains increase therapeutic efficacy by delivering multiple high-LET emissions, they also present challenges in controlling the spatial distribution of dose. Daughter isotopes can differ in energy, half-life and chemical behaviour, and their recoil-induced release from the original targeting vector can lead to off-target toxicity. To address this, researchers are developing strategies such as encapsulation in nanoparticles, use of *in vivo* scavengers and advanced chelators that attempt to retain progeny near the treatment site.

The physics of dose delivery in alpha therapy differs fundamentally from beta-emitting treatments. Because energy is deposited over subcellular distances, conventional macroscopic dosimetry models are insufficient. Instead, track-structure simulations using Monte Carlo codes such as GEANT4-DNA or MCNP are employed. GEANT4-DNA, in particular, includes physics models tailored to interactions at the nanometre scale, allowing simulations of energy deposition within cell nuclei or even DNA strands. These models require accurate inputs including alpha-particle energies, stopping powers, angular distributions and decay kinematics.

Producing alpha emitters for medical use remains a major bottleneck. ^{223}Ra is typically extracted from ^{227}Ac generators or produced by neutron irradiation of ^{226}Ra. ^{225}Ac can be obtained through separation from ^{233}U decay chains or by spallation of thorium targets using high-energy protons, followed by radiochemical purification. Each method involves trade-offs in cost, yield and radionuclidic purity, and requires precise optimization of irradiation and separation conditions.

Due to the high energy per decay and the potential for redistribution of daughters, alpha therapy necessitates stringent radioprotection protocols. These include shielding, controlled handling, and secure transport and waste management procedures, which are guided by international standards such as those from the International Atomic Energy Agency (IAEA) and the International Commission on Radiological Protection (ICRP). As TAT continues to gain clinical relevance, especially with the emergence of new theranostic pairs and improved targeting strategies, its advancement depends on close integration of nuclear physics, radiochemistry and biology.

6.1.3.3 Auger therapy
A third class of emitters includes radionuclides that release Auger electrons or internal conversion electrons, such as ^{111}In, ^{161}Tb and ^{125}I. These low-energy electrons travel only subcellular distances, typically less than a micron, and require localization near the cell nucleus to induce lethal DNA damage.

6.1.4 Theranostics

Theranostics is a strategy in nuclear medicine that uses radiopharmaceuticals to both visualize *and* treat disease in a coordinated way. A diagnostic radiotracer is first administered to track the distribution of a compound in the body using PET or SPECT. If the uptake in tumours is favourable, a therapeutic analogue is given, often based on the same chemical structure, to deliver a radiation dose to those same sites. Importantly, imaging can also be performed after therapy to assess response and confirm targeting. This sequential use of imaging and therapy enables treatment plans to be tailored to the individual, improving safety and effectiveness. Theranostic agents often share a common molecular framework for both functions, using the same chelator–vector combination but differing in the radioactive isotope attached.

In many cases, the imaging and therapeutic radionuclides belong to the same chemical element. These so-called *matched pairs*, such as $^{64/67}$Cu or $^{152/161}$Tb, share nearly identical coordination chemistry and biodistribution, so the same chelator–vector combination can be applied to both, ensuring consistent pharmacokinetics between imaging and therapy. However, theranostics can also be carried out using radionuclides of different elements—as long as the pharmacokinetics remain comparable—such as in the well-established combination of ^{68}Ga and ^{177}Lu [14].

A persistent challenge in internal radiotherapy is accurately determining how much radiation different tissues absorb, particularly when the therapeutic isotope does not emit photons suitable for imaging. For example, ^{131}I—used clinically for nearly 80 years—emits both β^- particles and gamma photons, allowing imaging with SPECT. However, the quantitative accuracy of this imaging for dosimetry is limited. The challenge is even greater for pure β^- emitters such as ^{90}Y and ^{89}Sr, which lack gamma emissions entirely and therefore cannot be imaged reliably using standard gamma-based methods. Although techniques such as bremsstrahlung SPECT and internal pair production PET have been explored, their limited resolution and low quantitative accuracy make them suboptimal for dosimetric planning. These limitations have motivated the use of matched diagnostic isotopes, which serve primarily not for imaging but rather for quantitative dosimetry. By mimicking the biodistribution of the therapeutic agent, these isotopes allow more reliable pre-treatment planning and post-treatment dose verification.

A striking example of a theranostic pair is $^{203/212}$Pb. The imaging isotope ^{203}Pb emits gamma photons suitable for SPECT, while the therapeutic ^{212}Pb decays via β^- emission to ^{212}Bi, which itself undergoes both β^- and high-LET α decay. This chain of emissions allows TAT to be guided and monitored using a chemically identical diagnostic agent—particularly valuable because α particles cannot be detected directly *in vivo*. Preclinical studies in prostate cancer models have shown that this pair enables both effective tumour imaging and therapeutic response, highlighting its promise for image-guided alpha therapy [15].

Table 6.1 presents selected radionuclide pairs that have been studied or applied in theranostics. Some, such as ^{68}Ga/^{177}Lu and ^{124}I/^{131}I, are routinely used in clinical

Table 6.1. Decay characteristics of selected theranostic radionuclide pairs, including half-lives, decay modes, mean β and γ emission energies, and their respective intensities. Imaging isotopes are indicated in rose, and therapeutic isotopes in blue. Nuclear data are taken from NuDat. For ^{212}Pb, which acts as an *in vivo* generator of α-emitting ^{212}Bi, the full decay chain and energy branching are shown in the footnote.

Nuclide	Half-life	Decay mode	$E_{\beta_{mean}}$ (keV)	I_β (%)	E_γ (keV)	I_γ (%)
^{43}Sc	3.891 h	β^+/EC	476	88.1	372.9	22.5
^{44}Sc	4.0420 h	β^+/EC	630.2	94.278	1157.0	99.88
^{47}Sc	3.3492 d	β^-	162.0	100	159.38	68.3
^{64}Cu	12.700 h	β^+/EC; β^-	278; 191	17.49; 38.5	1345.77	0.472
^{67}Cu	61.83 h	β^-	141	100	184.5, 93.3	48.7, 16.1
^{83}Sr	32.41 h	β^+/EC	510	27	762.6, 381.5	26.7, 14.0
^{89}Sr	50.563 d	β^-	587.1	100	—	—
^{86}Y	14.74 h	β^+/EC	660	31.9	1076.6, 627.7	82.5, 32.6
^{90}Y	64.05 h	β^-	932.3	100	—	—
^{124}I	4.1760 d	β^+/EC	820	22.7	602.7, 722.7	62.9, 10.3
^{131}I	8.0252 d	β^-	181.9	100	364.4, 636.9	81.5, 7.1
^{152}Tb	17.5 h	β^+/EC	1140	20.3	344.2, 271.0	63.5, 9.5
^{161}Tb	6.89 d	β^-	154	100	48.9, 74.5	17.0, 10.2
^{149}Tb	4.12 h	α; β^+	3967[i]; 720	16.7[ii]; 7.11	352.2, 164.9	29.8, 26.7
^{68}Ga	67.71 m	β^+/EC	829.5	88.91	1077.3	3.22
^{177}Lu	6.6443 d	β^-	133.6	100	208.3, 112.9	10.41, 6.23
^{203}Pb	51.92 h	EC	—	—	279.1	80.94
^{212}Pb[iii]	10.622 h	$\beta^- \rightarrow {}^{212}$Bi ($\alpha + \beta^-$)	101.3	100	238.6, 300.0	43.6, 3.30

[i] E_α (keV).
[ii] I_α (%).
[iii] Decay chain of ^{212}Pb:

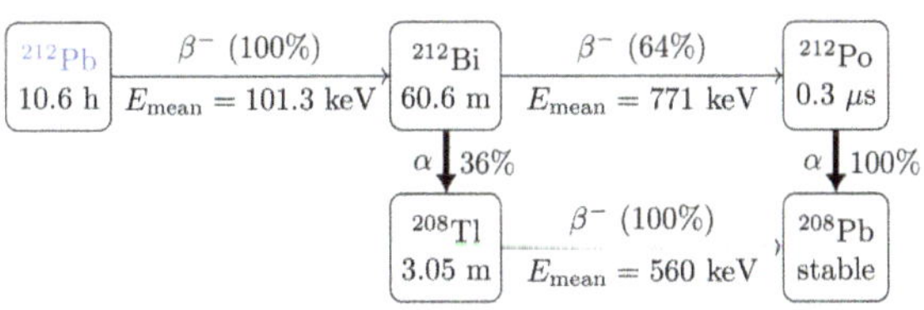

practice. Others, including terbium and lead-based pairs, are at various stages of preclinical and early clinical evaluation. The range of decay types and emission properties in these examples reflects the flexibility and expanding role of theranostics in personalized cancer care.

6.2 Experimental methods for nuclear reaction data

Reliable nuclear data require well-controlled irradiations, accurate quantification of reaction products and a thorough understanding of the experimental conditions. This section introduces the principal tools and techniques used to measure high-quality reaction cross sections, with particular emphasis on the beam facilities, target design and detection methods relevant to the case studies that follow.

While neutron-induced reactions—central to the operation of fusion and fission reactors—were discussed in earlier chapters, the present focus is on charged-particle- and photon-induced reactions. These reactions underpin the majority of the routes used for production of medical radioisotopes. The experiments described here primarily involve proton beams from cyclotrons and bremsstrahlung photons from electron linacs, but we also highlight how high-energy proton beams at facilities such as CERN's ISOLDE facility contribute to nuclear data through the production of rare isotopes for decay spectroscopy. Together, these methods form a versatile experimental framework for characterizing reaction cross sections and refining the nuclear data libraries essential to nuclear medicine.

6.2.1 Beam sources

6.2.1.1 Cyclotrons

Compact cyclotrons are widely employed in medical isotope production and nuclear reaction studies. These accelerators use a fixed magnetic field and alternating electric potential to accelerate charged particles—typically protons, deuterons or alpha particles—along outward spiral paths. Most cyclotrons dedicated to radioisotope production or cross section measurements operate at final energies between 10 and 30 MeV, although systems reaching 70 MeV or higher are available at some research facilities. Beam currents range from hundreds of nanoamperes for nuclear data measurements to hundreds of microamperes in high-throughput isotope production.

In experimental nuclear data work, cyclotrons support precise measurement of reaction cross sections using techniques such as stacked-foil activation. Their tunable energy and current make them well suited for studying excitation functions and quantifying yields in light-ion-induced reactions. Many production cyclotrons are also used for experimental campaigns, enabling continuity between data generation and applied isotope production.

6.2.1.2 Electron linacs

Electron linear accelerators (linacs) produce high-energy electron beams by accelerating electrons through a series of radio-frequency cavities. When directed onto a high-Z converter, typically tungsten or tantalum, these electrons undergo rapid deceleration in the nuclear Coulomb field and emit bremsstrahlung photons.

The resulting photon field spans a continuous energy spectrum up to the incident electron energy and is well suited to driving photonuclear reactions. This approach enables isotope production routes that are inaccessible or inefficient via direct charged-particle reactions. A prominent example is the $^{68}Zn(\gamma,p)^{67}Cu$ reaction, which circumvents the need for enriched nickel targets or high-current proton beams.

Electron linacs cover a wide energy range. Compact medical and industrial linacs typically operate between 10 and 50 MeV, while dedicated research installations, such as Mainz Microtron (MAMI) and Jefferson Lab, achieve several hundred MeV. Both continuous-wave and pulsed operation modes are available. The pulsed structure allows for coincidence measurements, time-of-flight techniques and suppression of uncorrelated background, which are particularly useful in photonuclear cross section studies.

An advanced capability offered by some electron accelerator facilities is photon tagging. In this method, the energy of each bremsstrahlung photon is determined by measuring the momentum of the scattered electron using a magnetic spectrometer. This produces a quasi-monoenergetic tagged photon beam with known energy resolution and timing characteristics. Tagged beams enable differential cross section measurements with well-defined incident energies, providing crucial benchmarks for photonuclear data and nuclear structure studies.

Electron linacs also deliver high instantaneous photon rates and can be designed for compact, modular configurations. These features make them attractive platforms for both experimental nuclear data measurements and emerging accelerator-based radioisotope production schemes.

6.2.1.3 High-energy proton sources

Some of the most precise decay data for medical radionuclides originate from facilities that employ high-energy proton beams to produce rare isotopes. At CERN's ISOLDE facility, 1.4 GeV protons from the Proton Synchrotron Booster (PSB) are delivered to thick, high-Z targets. Interactions such as spallation, fragmentation and fission produce a broad range of neutron-rich and proton-rich nuclei. These products are ionized, mass separated and delivered as low-energy beams for high-precision decay spectroscopy. The resulting measurements of half-lives, branching ratios and gamma-ray emission probabilities provide critical input for nuclear medicine dosimetry, decay heat calculations and quality assurance.

A complementary programme operates at TRIUMF in Canada, where a 500 MeV cyclotron delivers high-current protons to a range of dedicated target stations. Certain alpha-emitting radionuclides, such as ^{225}Ac, are extracted from the beam dump via spallation of thorium metal [16]. These efforts support the development of emerging radioisotopes and improve their associated decay data for therapeutic applications.

High-energy proton facilities thus occupy a specialized but essential position in the nuclear data infrastructure, enabling decay characterization of radionuclides that are otherwise inaccessible through conventional production routes.

6.2.2 Target preparation and energy loss

Accurate nuclear reaction measurements begin with well-defined and thoroughly characterized targets. Depending on the isotope and reaction channel under study, targets may be prepared as metallic foils, electroplated layers, pressed powders or sealed cells containing gaseous or liquid materials. Each form presents a trade-off between mechanical stability, thermal performance, chemical reactivity and post-irradiation handling. Metallic foils and electroplated films are widely used in stacked-target experiments due to their uniform thickness and good heat conduction. Gas and liquid targets, while more complex to fabricate and operate, are essential for certain production routes, such as the production of ^{18}F from ^{18}O-enriched water.

Isotopic enrichment is often required to enhance production yield and suppress the formation of unwanted radionuclidic impurities. Although enriched materials improve specificity, they are costly and may be in limited supply. These constraints must be carefully considered when selecting reaction pathways and designing target systems.

A key aspect of experimental planning is the accurate determination of the beam energy E at each depth x within a stacked target. As charged particles penetrate matter, they lose energy primarily through interactions with atomic electrons, resulting in ionization and excitation. This process is described by the stopping power, $S(E) = -\mathrm{d}E/\mathrm{d}x$, which quantifies the energy lost per unit path length as a function of the particle's energy. In stacked-foil arrangements, knowledge of $S(E)$ allows the initial beam energy to be mapped onto each target layer, providing the incident energy at which the reaction cross section is measured. This is particularly important for reactions near thresholds or rapidly changing excitation functions.

Stopping powers may be derived within the Bethe–Bloch formalism which is outlined in more detail in appendix A. In practice, stopping powers are readily obtained from well-established computational tools rather than calculated *ab initio*. The SRIM code provides tabulated values for a wide range of ions and materials based on semi-empirical models and fits to experimental data. Transport codes such as GEANT4 and FLUKA incorporate stopping power formulations that are benchmarked against reference data from the ICRU Report 49 and ICRU Report 90, as well as the NIST stopping power and range tables, particularly for protons, alpha particles and electrons. These tools ensure compatibility with international standards, in particular in applications related to medical dosimetry and radiological protection.

6.2.3 Irradiation and beam monitoring

During irradiation, the beam current and exposure time are recorded to determine the incident particle fluence—the total number of particles crossing a unit area. Accurate fluence determination is essential for calculating reaction cross sections from activation data, as it sets the absolute scale for measured yields.

6.2.3.1 Faraday-cup charge integration

For single-foil irradiations or bulk production runs, the total incident charge is often measured using a Faraday cup or current integrator placed downstream of the target. The collected charge is converted to particle fluence using the known charge state of the incident beam and the beam spot area. To ensure accuracy, the charge collection system must be properly calibrated, including corrections for secondary-electron emission, backscattering and any beam losses outside the collection geometry. When these effects are controlled, fluence uncertainties can typically be kept below 5%. In set-ups lacking a dedicated Faraday cup, a simplified configuration—sometimes called a 'poor man's Faraday cup'—may be used. This typically involves measuring current from the beam striking the target holder or a rear beam stop, with the structure grounded and instrumented to collect charge. While convenient, this method is more susceptible to systematic errors due to uncollected charge, electron escape and undefined beam geometry. Careful benchmarking against standard charge integration methods is therefore required if it is to be used for quantitative cross section work.

6.2.3.2 Flux monitors in stacked targets

In stacked-target experiments, the beam energy and particle fluence vary with depth due to energy loss and attenuation. To map this variation, thin *monitor* foils of known composition—typically copper, aluminium or titanium—are interleaved with the target foils. Each monitor is irradiated under the same conditions as the target layers, then measured using gamma-ray spectrometry. After correcting the observed activity to the end of bombardment (EOB), the local fluence is extracted by comparison with the recommended cross section of the selected monitor reaction at that energy.

Figure 6.1 shows a selection of IAEA-evaluated monitor reactions and their uncertainties. The Padé-fit curves are compact functional representations of the

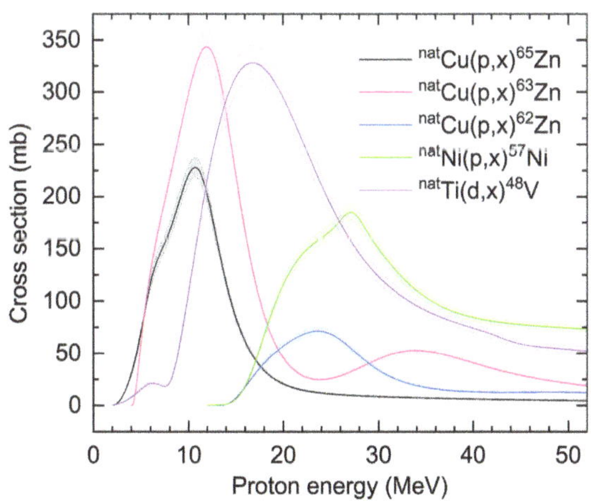

Figure 6.1. IAEA-recommended monitor reactions for charged-particle cross section measurements. Curves show Padé fits to experimental data; shaded regions represent the evaluated uncertainty [18]. The full set of monitor data is available via the IAEA Medical Charged-Particle Cross section Database, covering proton-, deuteron-, ^{3}He- and ^{4}He-induced reactions up to 1 GeV.

evaluated cross section data, typically expressed as a rational function or modified empirical form optimized for nuclear data applications [18]. These fits are applied post-evaluation to facilitate smooth interpolation, numerical integration and practical implementation across a variety of reaction types, including monitor reactions. Examples of such Padé-type representations, along with fitting parameters, are provided in the IAEA Medical Charged-Particle Cross section Database [18]. In practice, a monitor reaction is chosen to match the energy range of interest, ensuring both overlap with the target reaction and acceptable uncertainty. During analysis, the monitor activities are used to construct an energy-fluence profile, which is then applied to the neighbouring target foils.

Because the beam energy degrades across the stack, using multiple monitor reactions with different thresholds and peak cross sections improves confidence in the assigned energy-fluence grid. This redundancy reduces systematic uncertainties and increases the robustness of the extracted excitation function, particularly in regions of steep cross section variation. A typical workflow is as follows:

1. Estimate the proton energy at each monitor position using stopping power calculations based on the initial beam energy and the sequence of material layers in the stack.
2. Measure the activity of each monitor foil using gamma-ray spectrometry. Correct the recorded counts for detector efficiency, decay after irradiation during the cooling and counting periods, and any dead time effects to obtain the EOB activity A_{EOB}.
3. Obtain the recommended cross section $\sigma(E)$ and its uncertainty for the monitor reaction at the estimated energy, using evaluated datasets such as the IAEA Medical Charged-Particle Cross Section Database.
4. Calculate the local particle fluence $\phi(E)$ using

$$\phi(E) = \frac{A_{\text{EOB}}}{\sigma(E)\, n\, (1 - e^{-\lambda t_i})}, \tag{6.1}$$

   where n is the areal density of target atoms, given by $n = \frac{N_A \rho d}{M}$, with N_A as Avogadro's number, ρ the material density, d the target thickness and M the molar mass. Here, λ is the decay constant of the activation product and t_i is the irradiation time.
5. Assign the calculated fluence to the main target foils adjacent to each monitor. This yields a depth-resolved energy-fluence map across the stack, forming the basis for cross section determination.

6.2.4 Gamma-ray spectrometry

Each activated foil is assayed following irradiation using calibrated high-purity germanium (HPGe) detectors. After a short cooling period to allow short-lived activity to decay, the net photopeak area C is extracted from the gamma spectrum and corrected for detector efficiency, decay, geometry and dead time to determine A_{EOB} as

$$A_{\mathrm{EOB}} = \frac{C\,\lambda}{1 - e^{-\lambda t_{m_r}}} \cdot \frac{1}{e^{-\lambda t_c}} \cdot \frac{t_{m_r}}{t_{m_l}} \cdot \frac{f_a}{\varepsilon\, I_\gamma}, \tag{6.2}$$

where t_c is the cooling time, t_{m_r} and t_{m_l} are the real and live counting times, ε is the full-energy peak efficiency, I_γ is the gamma emission probability and f_a accounts for self-attenuation within the foil. The attenuation correction is given by

$$f_a = \frac{(\mu/\rho)\,\rho d}{1 - e^{-(\mu/\rho)\,\rho d}}, \tag{6.3}$$

where μ/ρ is the mass attenuation coefficient. Coefficients for a wide range of materials and photon energies are available from the NIST XCOM database.

The resulting A_{EOB} is used to determine the particle fluence in monitor foils and, in turn, to derive reaction cross sections in the surrounding target layers.

6.2.5 Cross sections and yields

Cross sections are calculated from the measured A_{EOB} by rearranging equation (6.1):

$$\sigma(E) = \frac{A_{\mathrm{EOB}}}{\phi(E)\, n\, (1 - e^{-\lambda t_i})}, \tag{6.4}$$

where all terms have been previously defined.

For thick targets where the projectile energy degrades significantly within the material, the physical yield, defined as the total A_{EOB} per unit beam charge, is obtained by integrating the cross section over the energy loss:

$$Y(E_0) = \frac{N_A}{M} \int_{E_{\mathrm{th}}}^{E_0} \sigma(E)\, \frac{1}{S(E)}\, \mathrm{d}E, \tag{6.5}$$

where E_0 is the incident energy, E_{th} is the reaction threshold and $S(E)$ is the stopping power of the target material. The yield is typically expressed in units of MBq μA^{-1} h^{-1} and guides operational planning in isotope production, including beam time, target design and activity forecasting. Note the similarity in approach here to that for (α,n) reactions in spent fuel described in section 5.3.1, for example, compare equation (6.5) with equation (5.6).

With the experimental framework established, we now turn to the first of two production case studies.

6.3 Case study 1: ^{67}Cu production

6.3.1 Theranostic role of ^{67}Cu

Copper provides a unique family of radioisotopes that span both diagnostic and therapeutic functions, making it highly valuable for nuclear medicine. As summarized in table 6.2, positron-emitting isotopes such as ^{60}Cu, ^{61}Cu, ^{62}Cu and ^{64}Cu are well suited for PET imaging, while ^{64}Cu and ^{67}Cu offer therapeutic potential through β^- emission. The pairing of ^{64}Cu for PET and ^{67}Cu for therapy forms an attractive theranostic combination that allows the same chemical vector to be used

Table 6.2. Specific decay properties of copper isotopes, including the decay mode, mean beta and γ-ray energies and their principal intensities.

Isotope	Half-life	Decay mode	$E_{\beta_{mean}}$ (keV)	I_β (%)	E_γ (keV)	I_γ (%)
^{60}Cu	23.7 m	β^+/EC	970	93	826.4	21.7
					1332.5	88.0
					1791.6	45.4
^{61}Cu	3.339 h	β^+/EC	500	61	282.956	12.7
					656.008	10.4
^{62}Cu	9.67 m	β^+/EC	1319	97.83	875.66	0.147
					1172.97	0.342
^{64}Cu	12.700 h	β^+/EC	278	17.49	1345.77	0.472
		β^-	191	38.5	–	–
^{66}Cu	5.12 m	β^-	1066	100	1039.2	9.23
					833.0	0.220
^{67}Cu	61.83 h	β^-	141	100	91.266	7.00
					93.311	16.10
					184.577	48.7
					300.219	0.797

for both diagnostic imaging and treatment. In some cases ^{61}Cu can serve as the diagnostic counterpart, although its shorter half-life and higher positron energy make it less ideal for quantitative imaging.

Importantly, ^{67}Cu can also act as a standalone theranostic agent via SPECT imaging, thanks to its accompanying gamma emissions. With a half-life of 61.83 h, it decays exclusively by β^- emission with a mean energy of 141 keV, accompanied by photons at 91.2, 93.3 and 184.5 keV, as shown in figure 6.2. This radiation profile is well matched to treating small tumours or micrometastases while permitting *in vivo* tracking through gamma detection. The intermediate half-life enables centralized production and regional distribution, with sufficient time for radiolabelling and clinical use yet short enough to limit long-term radioactive waste.

Clinical adoption of ^{67}Cu has lagged behind its promise because production is technically demanding. In contrast to ^{64}Cu, which is readily produced on low-energy cyclotrons, ^{67}Cu requires enriched targets and careful tuning of beam energy to minimize isotopic impurities.

6.3.2 Production routes

There are several different nuclear reactions that can be used to produce ^{67}Cu, each with its own advantages and drawbacks. Table 6.3 summarizes the main production routes, listing the energy thresholds and typical energy ranges needed for each reaction. Figure 6.3 shows the theoretical cross sections for these reactions as calculated using the code TALYS [19]. These curves help identify the energy ranges where each reaction is most effective.

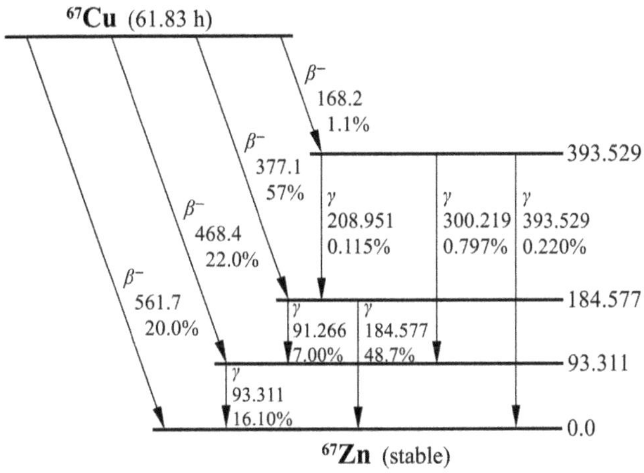

Figure 6.2. Decay scheme for the β^- decay of ^{67}Cu into ^{67}Zn. Beta transitions are labelled with their respective endpoint energy in keV and their branching ratio in %. Subsequent gamma-ray transitions are labelled with their transition energy in keV and their branching ratio in %.

Table 6.3. Nuclear reactions for the production of ^{67}Cu, with corresponding Q-values, threshold energies E_{th} and approximate effective energy ranges for production. The listed energy ranges were estimated from TALYS-calculated excitation functions (see figure 6.3) by identifying the main peak regions, without accounting for impurity production or full yield integration.

Reaction	Q-value (MeV)	E_{th} (MeV)	Typical production energy range (MeV)
$^{68}Zn(\gamma,p)$	−9.97	9.97	15–25
$^{70}Zn(p,\alpha)$	2.61	0	10–25
$^{70}Zn(p,2n2p)$	−25.67	26.04	>50
$^{68}Zn(p,2p)$	−9.97	10.12	>35
$^{64}Ni(\alpha,p)$	−4.64	4.89	15–35
$^{70}Zn(d,n\alpha)$	0.39	0	15–30
$^{67}Zn(n,p)$	0.22	0	<40

Proton- and deuteron-induced reactions on enriched zinc targets, particularly $^{70}Zn(p,\alpha)^{67}Cu$ and $^{70}Zn(d,n\alpha)^{67}Cu$, represent well-established production routes for ^{67}Cu. These reactions are accessible using low- to medium-energy cyclotrons commonly employed in medical isotope production, making them operationally compatible with existing infrastructure. However, the low natural abundance of ^{70}Zn (0.61%) necessitates the use of isotopically enriched targets, which significantly increases material costs. Despite this limitation, these reactions offer a favourable radionuclidic purity, yielding ^{67}Cu with minimal co-production of isotopic contaminants such as ^{64}Cu.

In contrast, high-energy proton-induced reactions such as $^{68}Zn(p,2p)^{67}Cu$ can achieve higher yields due to the increased reaction cross section at elevated energies.

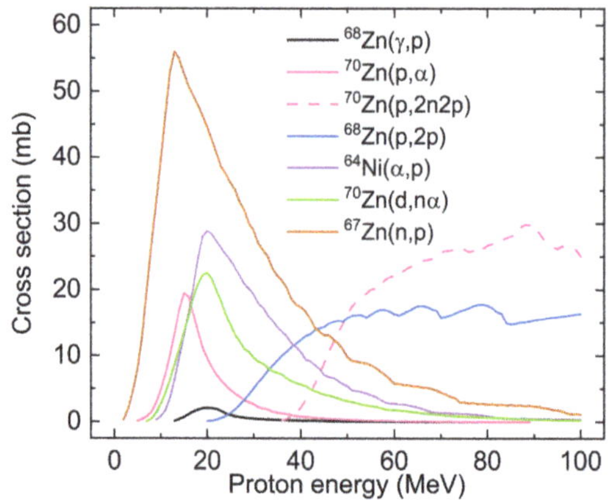

Figure 6.3. Excitation functions for selected nuclear reactions producing ^{67}Cu, calculated using TALYS. The plots cover reactions induced by protons, deuterons, alpha particles, photons and neutrons.

However, access to such high-energy cyclotrons is limited, and this route suffers from reduced radionuclidic purity, as it produces substantial quantities of ^{64}Cu alongside the desired product. Since both isotopes are chemically indistinguishable, post-irradiation separation is not feasible, resulting in a reduced specific activity. This compromises therapeutic applications, as a lower specific activity means fewer radioactive atoms are delivered per targeting molecule, reducing tumour dose while allowing non-radioactive carriers to block receptor sites.

An alternative and increasingly popular method involves using high-energy photons from an electron accelerator. The reaction $^{68}Zn(\gamma,p)^{67}Cu$ begins at about 10 MeV and peaks between 15 and 25 MeV, as shown in figure 6.3. This method works well with both natural and enriched ^{68}Zn targets and it produces very little ^{64}Cu, leading to high-purity ^{67}Cu. Photon-induced reactions also allow for bulk irradiation of multiple foils or large-area targets simultaneously, which helps with scaling up production. The main drawback is that relatively few facilities currently operate high-current electron linacs, although this is gradually changing with advances in accelerator technology, including emerging developments in compact systems such as laser-plasma acceleration [20, 21], which can generate high-energy electron beams using ultra-intense laser pulses on thin foil targets.

Neutron-induced production routes have also been explored, such as the $^{67}Zn(n,p)^{67}Cu$ reaction. This process can be performed in research reactors or accelerator-driven neutron sources capable of producing fast or quasi-monoenergetic neutrons. However, it requires the use of enriched ^{67}Zn targets and tends to produce impurities such as ^{64}Cu, which compromises the radionuclidic purity. In addition, the need for specialized neutron fields limits the practicality of this method, making it less suitable for routine or large-scale production.

Only a few facilities around the world are currently able to produce clinical-grade ^{67}Cu reliably. For example, the Idaho Accelerator Center uses the photonuclear route to produce multi-gigabecquerel batches. Smaller quantities are also produced at research centres such as the Hevesy Laboratory in Denmark and ARRONAX in

France, often for preclinical studies and method development. These facilities support initiatives such as PRISMAP, which fosters early-stage radiopharmaceutical research and advances nuclear medicine applications.

In practice, the choice of production route depends on a number of factors: the desired radionuclidic purity, the availability of enriched targets, the beam energy available at a given facility and the economic feasibility of large-scale production. Cross section data, such as those shown in figure 6.3, help guide these decisions by identifying suitable energy ranges for maximizing yield while minimizing impurities. However, theoretical predictions are not always accurate near threshold or in reactions involving complex mechanisms or competing channels, so experimental measurements remain essential. In the next sections, we present two case studies that explore the measurement and optimization of ^{67}Cu production via two routes: one using protons on ^{70}Zn and another using bremsstrahlung photons on ^{68}Zn.

6.3.3 Proton-induced route: ^{70}Zn(p,α)

Accurate cross section measurements are foundational for evaluating the suitability of a reaction for medical isotope production. For ^{67}Cu, the ^{70}Zn(p,α)^{67}Cu reaction offers one of the most promising low-energy charged-particle routes, particularly suitable for medical cyclotrons. The experimental process begins with stacked-target activation, a technique that enables measurement of excitation functions across a range of incident proton energies in a single irradiation.

A typical stacked target, illustrated in figure 6.4, consists of alternating layers of thin zinc foils, aluminium degraders and copper monitor foils. The zinc layers—often enriched ^{70}Zn, although in some cases natural zinc may be used if the reaction channels remain distinguishable—serve as the primary target material. As the incident proton beam penetrates the stack, it gradually loses energy due to interactions with the aluminium degraders and intervening foils. This energy degradation ensures that each zinc foil is irradiated at a slightly lower proton energy

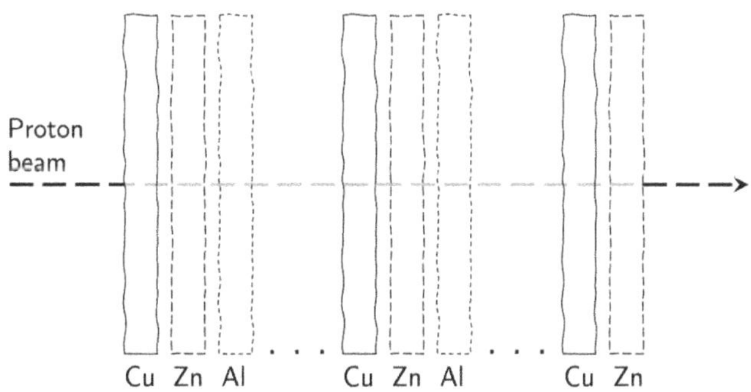

Figure 6.4. Schematic of a stacked-target arrangement used for cross section measurements. A proton beam of defined initial energy enters the stack, which is composed of repeated layers of copper (flux monitor), zinc (main target) and aluminium (energy degrader) foils. As the protons pass through the stack, their energy decreases progressively. The beam flux is monitored using copper foils via the ^{nat}Cu(p,x)^{62,63,65}Zn reactions, with the proton energy at each foil position determined through stopping power calculations.

than the one before, effectively sampling the excitation function across a broad energy range in a single irradiation.

Copper foils embedded between the layers act as beam monitors, where well-characterized $^{nat}Cu(p,x)^{62,63,65}Zn$ reactions are used to reconstruct the local proton fluence and energy at each position (see section 6.2.3). This is achieved by measuring the induced activities via gamma-ray spectrometry and converting them to fluence using IAEA-recommended monitor cross sections. The energy associated with each foil is determined through stopping power calculations (see section 6.2.2).

After target assembly, the stacked foils are irradiated with a proton beam at a facility capable of delivering suitable energies for charged-particle activation. For this case study, we describe an experiment carried out at the Birmingham Cyclotron [22], where proton beams of up to 40 MeV are available. This energy range comfortably spans the threshold and peak regions of the $^{70}Zn(p,\alpha)^{67}Cu$ reaction, making it ideal for mapping its excitation function. During irradiation, the proton beam is directed onto the upstream face of the stacked target, typically for several minutes to a few hours. Beam currents are kept low—on the order of a few hundred nanoamperes—to minimize activation and radioactive waste, as is common practice in cross section studies. This approach balances the need for sufficient counting statistics with the aim of limiting target activity for safe handling and waste management.

Following irradiation, the foils are carefully disassembled and individually analysed by high-resolution gamma-ray spectrometry. This allows the identification and quantification of ^{67}Cu and other products generated in each foil. The 184.6 keV gamma emission from ^{67}Cu is especially useful due to its relatively strong branching ratio and minimal interference from other lines. Figure 6.5 shows representative spectra from irradiated copper and zinc foils, illustrating characteristic peaks and the need for spectral deconvolution where isotopes such as ^{67}Ga are also produced.

By applying equation (6.2) to the photopeak areas, we obtain the A_{EOB} of ^{67}Cu in every foil. Combining these activities with the measured fluence, target areal density and irradiation time in equation (6.4) yields the experimental cross section for the mean proton energy in each layer. Plotting those data points produces an excitation function similar to that shown in figure 6.6, which presents two experimental datasets for this reaction alongside TALYS-calculated and IAEA-recommended curves. The cross section rises steeply above threshold, reaches a maximum of approximately 15 mb near 15 MeV, and then gradually decreases at higher energies. This profile defines a practical irradiation window of roughly 10–25 MeV for optimal production of ^{67}Cu.

Integrating the measured cross section over the energy lost in the target (equation (6.5)) converts the curve into the physical yield shown in figure 6.7. The yield increases through the effective energy range and begins to plateau above 30 MeV, indicating that higher beam energies do not significantly enhance production. A practical irradiation window from about 25 MeV down to threshold captures the majority of the achievable ^{67}Cu yield. Operating much above this window is undesirable because impurity production becomes significant. Figure 6.8 compares excitation functions for other $^{70}Zn(p,x)$ channels: reactions leading to ^{64}Cu or

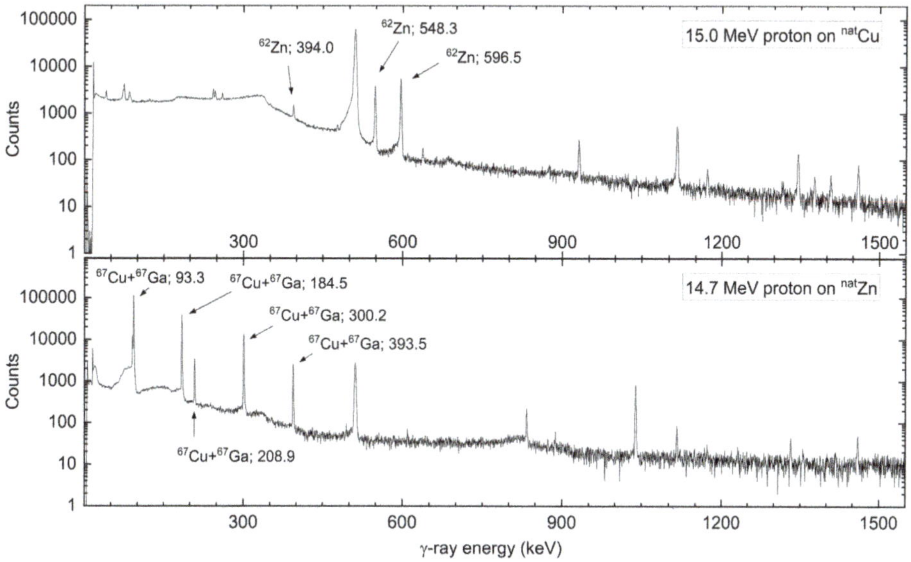

Figure 6.5. Representative gamma-ray spectra from foils irradiated in the stacked-target configuration shown in figure 6.4. Top: Spectrum from a natural copper foil, used as a monitor, showing prominent peaks from ^{62}Zn produced via the ^{nat}Cu(p,x)^{62}Zn reaction. These peaks are used to determine beam fluence using IAEA-recommended cross section standards (see figure 6.1). Bottom: Spectrum from a zinc foil positioned downstream of the copper monitor. Peaks from both ^{67}Cu and co-produced ^{67}Ga are labelled. Their contributions can be disentangled through decay curve fitting and gamma branching analysis to derive cross sections for the ^{nat}Zn(p,x)^{67}Cu and ^{nat}Zn(p,x)^{67}Ga reactions. Proton energies for each foil, calculated using SRIM, are indicated, with gamma peaks labelled in keV.

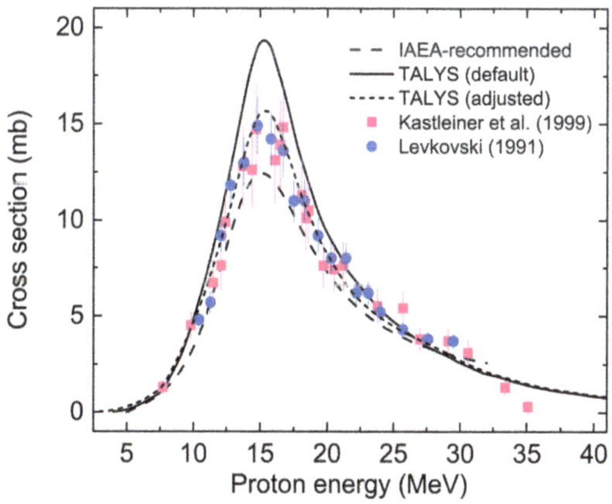

Figure 6.6. Excitation function of the ^{70}Zn(p,α)^{67}Cu reaction. Experimental data [23, 24] compiled from EXFOR are compared to IAEA-recommended and TALYS-calculated cross sections. The latter include default and adjusted models.

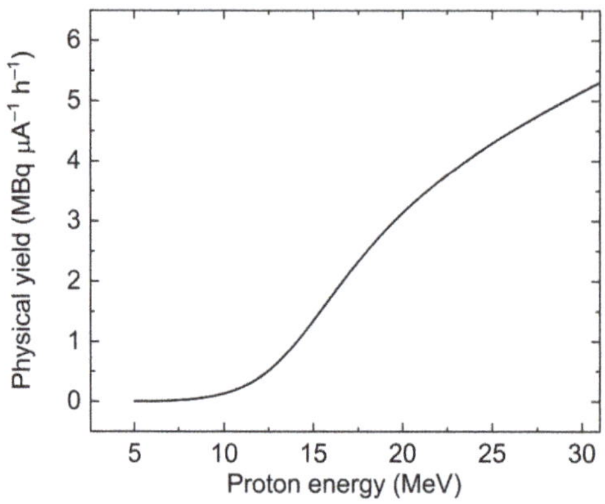

Figure 6.7. Physical yield for the $^{70}Zn(p,\alpha)^{67}Cu$ reaction, based on IAEA-recommended cross section data in figure 6.6.

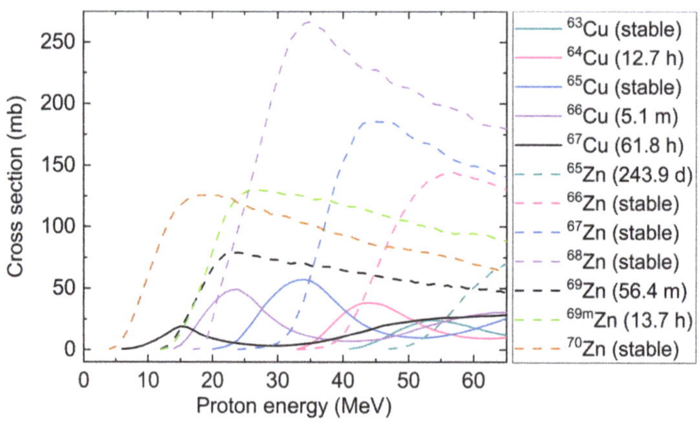

Figure 6.8. Excitation functions for $^{70}Zn(p,x)$ reactions leading to copper and zinc products. To minimize contamination from isotopic impurities, in particular stable ^{65}Cu, proton energies below 25 MeV are preferred. Zinc products can be chemically removed.

stable ^{65}Cu rise steeply beyond 25 MeV. Since those copper isotopes are chemically inseparable from ^{67}Cu, their presence dilutes the specific activity of the product. The impurity estimates in figure 6.8 are based on TALYS calculations; experimental cross section measurements for these channels remain sparse and would further refine production guidelines.

Altogether, this case study demonstrates how experimental cross section measurements guide energy selection, target design and yield estimation for medical isotope production. It also highlights the value of comparing experimental data with theoretical models such as TALYS to improve predictive power and inform future data library updates. The $^{70}Zn(p,\alpha)^{67}Cu$ route provides a practical example of low-energy cyclotron-based production that balances feasibility, purity and yield, particularly when enriched targets and careful energy planning are employed.

Complementing this approach is the photonuclear route, which avoids charged-particle beams altogether by using high-energy photons to induce the same nuclear transformation. The next case study investigates the experimental and theoretical aspects of ^{67}Cu production via the ^{68}Zn(γ,p) reaction.

6.3.4 Photonuclear route: ^{68}Zn(γ,p)

Unlike the previous case, where the experimental method was introduced first, we begin here by examining the available cross section data for the ^{68}Zn(γ,p)^{67}Cu reaction, as this informs both the theoretical understanding and the feasibility of photonuclear production. Figure 6.9 summarizes the current status of the experimental cross section information. All experimental data points derive from a single measurement conducted in the late 1950s [25], with sizeable uncertainties reported across the energy range.

To assess the theoretical spread, TALYS calculations are shown for six level density (LD) models (see description of level density in section 1.1.2). These include the default constant temperature plus Fermi gas model (LD1), the back shifted Fermi gas model (LD2), the generalized superfluid model (LD3) and three microscopic models based on Skyrme-HFB (LD4, LD5) and Gogny-HFB (LD6) calculations. A full description of the LD options and parameters can be found in [26]. While all LD variants predict an excitation function shaped by the giant dipole resonance (GDR), they differ significantly in magnitude. At certain energies, the highest predicted cross section exceeds the lowest by nearly a factor of four and most models underestimate the historical data.

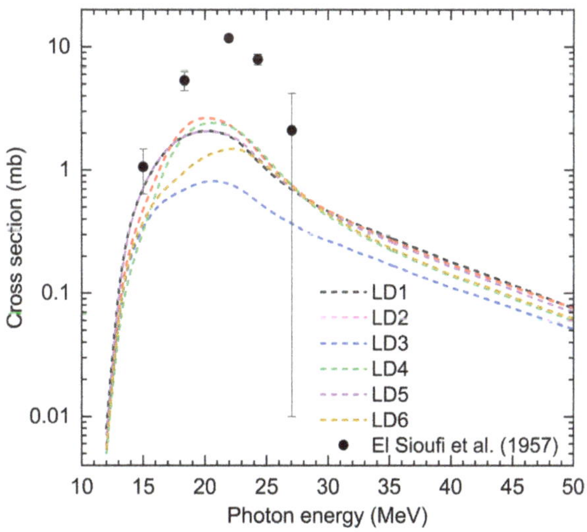

Figure 6.9. Experimental data for the ^{68}Zn(γ,p)^{67}Cu reaction compared with theoretical cross sections calculated using TALYS with different level density (LD) models (LD1–LD6). For a full description of the models, see [26]. The only available experimental data [25] exhibit significant uncertainty and are in partial disagreement with model predictions.

The strong sensitivity to level density model inputs highlights the critical role of nuclear structure assumptions in photonuclear cross section calculations. In the GDR region, the reaction mechanism involves photon absorption followed by competition between particle emission and statistical gamma de-excitation. The level density of the residual nucleus governs the probability of each decay channel. Without reliable experimental constraints on level densities, theoretical cross section predictions in this energy range remain uncertain.

Because the (γ,p) reaction on ^{68}Zn has a relatively low threshold around 10 MeV and peaks between 15–25 MeV, it is effectively driven by the bremsstrahlung photon spectra produced by electron linacs operating below 25 MeV. This has made the route attractive for isotope production at accelerator-based facilities. However, the absence of modern cross section measurements hampers the optimization of production parameters such as beam energy, target geometry and converter configuration.

The remainder of this section outlines an experimental strategy for determining reliable cross section data for the ^{68}Zn(γ,p)^{67}Cu reaction. The first step in this process is the production and characterization of the photon beam used to induce the reaction.

6.3.4.1 Photon beam set-up

The photon field that drives the ^{68}Zn(γ,p)^{67}Cu reaction must be accurately known in order to extract meaningful cross section data. This includes both the energy distribution of the photons and their fluence at the target position.

Photon sources

Bremsstrahlung linacs—A common method for generating high-energy photons is to direct a relativistic electron beam onto a dense metal target, often referred to as a radiator or converter. Materials with high atomic number, such as tungsten or tantalum, are typically used. As the electrons interact with the strong electromagnetic fields of the atomic nuclei in the converter, they are decelerated and emit bremsstrahlung radiation. This process produces a broad photon spectrum that extends from low energies up to the maximum kinetic energy of the incident electrons, known as the endpoint energy E_0. For thin targets, the emitted photon spectrum closely follows the characteristic shape of thin-target bremsstrahlung. However, in practice, the photon field is shaped by additional components in the beamline configuration, which influences the final photon spectrum through scattering and attenuation effects. Bremsstrahlung sources provide high photon intensities, in particular when operated at beam currents of several microamperes. Such set-ups can yield photon rates exceeding 10^{12} photons per second at the target position, which is more than adequate for activation experiments involving thin foils. This high intensity makes bremsstrahlung the workhorse of photonuclear data measurements, despite the fact that its continuous energy distribution complicates the extraction of pointwise cross section values.

Laser-Compton scattering—An alternative method for generating high-energy photons is laser-Compton scattering (LCS), used at facilities such as HIγS in the United States and NewSUBARU in Japan. In this approach, laser photons are scattered off relativistic electrons circulating in a storage ring. The inverse Compton interaction boosts the photon energy into the gamma-ray range, producing a beam that is quasi-monoenergetic, with typical energy resolutions between 2% and 5% and tunable photon energies up to about 100 MeV. Although the photon rates achieved by LCS systems are significantly lower than those from bremsstrahlung—typically in the range of 10^5 to 10^8 photons per second—the well-defined energy profile makes them highly valuable for nuclear data measurements. In particular, LCS beams provide precise pointwise cross section values that serve to benchmark or constrain results obtained from bremsstrahlung-based unfolding. A representative example of LCS generation using a storage-ring free-electron laser at HIγS is given in [27]. More recently, HIγS LCS beams have been used to measure cross sections for the ^{197}Au(γ,xn) reaction up to 31 MeV, extending the applicability of ^{197}Au as a photonuclear monitor standard [28]. This reaction will be discussed later in the context of its role as a reference in photon-induced activation experiments.

6.3.4.2 Photon spectrum and rate

Accurate description of the photon energy distribution at the target and its normalization is achieved by combining analytical models, Monte Carlo particle transport simulations and experimental calibration tools. The following approaches are typically employed.

Analytical estimate—The bremsstrahlung energy spectrum produced by an electron beam striking a thin converter can be approximated using the Schiff formula, which provides a convenient starting point for the differential photon flux $W(E_0, E_\gamma)$ as a function of incident electron energy E_0 and photon energy E_γ.

Monte Carlo modelling—To account for beamline geometry, material composition, collimation and photon attenuation, detailed simulations using codes such as GEANT4 and MCNP are performed. These simulations yield the differential photon flux $\phi(E)$ at the target location and can optionally resolve the spatial distribution across multiple foil positions in a stacked target. An example of such a simulated spectrum is shown in figure 6.10, where a GEANT4-generated distribution is compared with the theoretical Schiff spectrum.

Experimental normalization—To determine the absolute scale of the photon fluence, a reference monitor foil is irradiated alongside the main target. A thin gold (^{197}Au) foil is typically used, and its measured A_{EOB} is related to the total photon fluence Φ_γ via

$$\Phi_\gamma = \frac{A_{\mathrm{EOB}}}{n \left\langle \sigma_{\gamma,\,n}(E) \right\rangle (1 - e^{-\lambda t_i})}, \tag{6.6}$$

where n is the areal density of the gold target, t_i is the irradiation time, λ is the decay constant of the activation product and $\left\langle \sigma_{\gamma,\,n}(E) \right\rangle$ is the flux-weighted average cross section for the ^{197}Au(γ,n)^{196}Au reaction. Here, Φ_γ represents the total photon fluence, that is, the number of photons crossing a unit area over the entire irradiation, integrated across energy and time. It is related to the differential photon flux $\phi(E)$ via

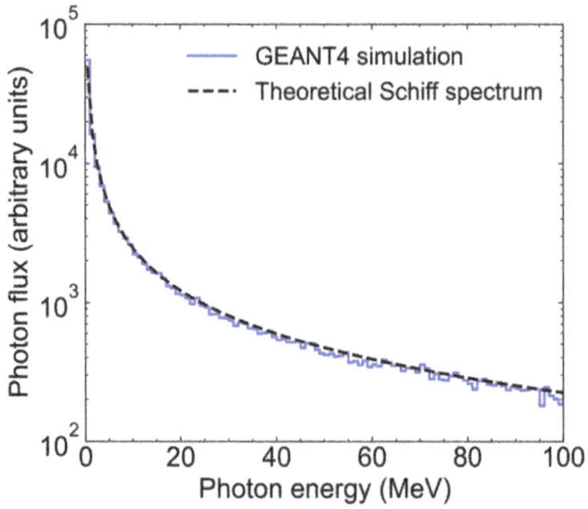

Figure 6.10. Comparison of a GEANT4-simulated bremsstrahlung photon energy spectrum with the analytical Schiff spectrum for a 100 MeV electron beam incident on a 10 μm copper converter. Both spectra are normalized to the same maximum value to allow direct visual comparison.

$$\Phi_\gamma = \int_0^{E_0} \phi(E)\, \mathrm{d}E. \tag{6.7}$$

This monitor reaction is well characterized across the relevant energy range and is recommended for bremsstrahlung normalization. An example of this approach is provided in [29], where gold and indium monitors were used to benchmark the photon fluence.

Spectral validation—Additional checks on the photon spectrum can be performed using detectors such as $LaBr_3$ or HPGe. While these are not typically used to directly measure the high-energy photon field due to limited efficiency and resolution above a few MeV, they can help verify the spectral shape below this range. Comparing the measured detector response with that folded from the simulated spectrum helps confirm the accuracy of the modelled bremsstrahlung low-energy contributions.

Tagged versus untagged photon beams—For photonuclear measurements, it is also useful to distinguish between untagged and tagged photon beams. Untagged bremsstrahlung beams deliver high flux and are suitable for activation experiments, but they do not provide event-by-event information on photon energy. In contrast, tagged photon systems use an electron spectrometer to determine the energy of each bremsstrahlung photon by detecting the scattered electron in coincidence. This technique allows for exclusive measurements with defined incident energies, often combined with particle detection in the exit channel and is particularly useful for differential cross section studies involving light nuclei or specific reaction products.

6.3.4.3 Cross section unfolding

When using bremsstrahlung photons, the measured activation yield at a given electron beam endpoint energy E_0 represents a spectrum-averaged quantity. The experimental observable is given by

$$Y(E_0) = \int_{E_{\mathrm{th}}}^{E_0} \sigma(E)\ \phi(E; E_0)\ \mathrm{d}E, \tag{6.8}$$

where $\sigma(E)$ is the energy-dependent reaction cross section, $\phi(E; E_0)$ is the differential photon flux at energy E for an electron beam with endpoint energy E_0 and E_{th} is the reaction threshold.

To extract the underlying cross section $\sigma(E)$ from a set of integrated yields, two complementary strategies are typically employed:

- **Differential endpoint method:** This approach involves performing a sequence of irradiations with increasing electron beam energies, for example at E_0, $E_0 + \Delta E$, $E_0 + 2\Delta E$ and so on. By taking the difference in yields between successive energies, the average cross section within each energy interval ΔE can be isolated. This method reconstructs $\sigma(E)$ in discrete slices and is especially effective when the photon flux increment between energy steps is well understood. Energy intervals of 2–3 MeV are a practical compromise between energy resolution and experimental effort.
- **Least-squares unfolding:** In this approach, the measured yield data from several beam energies are interpreted as a linear system:

$$\mathbf{Y} = \mathbf{K}\ \boldsymbol{\sigma},$$

where the kernel matrix **K** contains the integrated flux contributions in each energy bin and $\boldsymbol{\sigma}$ is the unknown vector of cross section values on a predefined energy grid, e.g. 1 MeV steps. The system is typically solved using non-negative least squares, with Tikhonov regularization applied to stabilize the solution and suppress unphysical fluctuations. While a theoretical cross section model from a code such as TALYS may be used to guide the initial binning or smoothing scale—often employing a Lorentzian parameterization of the GDR plus a quasideuteron tail—this does not strongly influence the result once sufficient data points are included in the fit.

The same deconvolution techniques can also be applied to LCS data, particularly when correcting for low-energy tails in the photon distribution. Although LCS sources are nominally quasi-monoenergetic, their spectra often display a sharp high-energy cut-off and an asymmetric tail toward lower energies, in particular at higher photon energies. In such cases, unfolding is used to recover the effective cross section. However, when the spectral width is narrow and well-characterized, a direct pointwise value $\sigma(E_\gamma)$ can be extracted without further deconvolution.

6.3.4.4 Production implications

In practical radioisotope production, such as routine ^{67}Cu manufacture, the emphasis shifts from precise cross section determination to maximizing total yield. To increase photon fluence, facilities typically use thicker converters—several millimetres in depth—and targets with thicknesses approaching two radiation lengths, corresponding to gram-scale masses. Beam currents are also scaled up significantly, often reaching tens to hundreds of microamperes, depending on the capabilities of the converter and target cooling systems. The resulting spectral broadening and loss of energy resolution are acceptable in a production context, where the goal is to maximize activity per unit time rather than resolve fine energy structures in the cross section. Nevertheless, the experimental methods described in the preceding sections remain essential. Accurate photonuclear cross section data are needed to optimize irradiation conditions, evaluate isotope purity and predict achievable yields. These nuclear data also form the foundation for economic and technical feasibility assessments of future accelerator-based production facilities using bremsstrahlung or other high-intensity photon sources.

While the production of ^{67}Cu illustrates the complexity and potential of both proton- and photon-induced reactions, many medical isotopes are routinely produced using simpler, well-established charged-particle reactions. Among these, ^{18}F remains the most widely used PET isotope worldwide. The next case study focuses on the production of ^{18}F, offering a contrast in both experimental approach and clinical application.

6.4 Case study 2: ^{18}F production

^{18}F is the most widely used positron emitter in nuclear medicine, serving as the backbone of PET imaging. Its most prominent clinical application is in ^{18}F-fluorodeoxyglucose (FDG), used to trace glucose metabolism in oncology, cardiology and neurology. The 109.77-minute half-life provides a practical window for multi-step radiolabelling, transportation and patient administration, while keeping radiation dose within manageable limits. As illustrated in figure 6.11, ^{18}F decays predominantly by positron emission (β^+, 96.73%), with an endpoint energy of 633.5 keV and a mean energy of 250 keV. This relatively low positron energy translates into a short range in tissue, improving image resolution compared to higher-energy emitters.

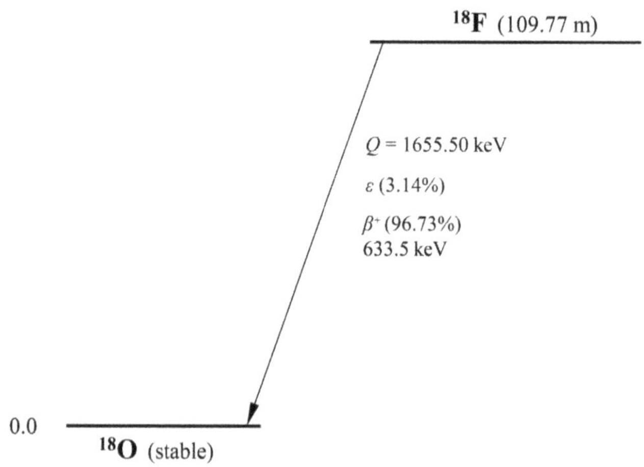

Figure 6.11. Decay scheme for the β^+ and electron capture ε decays of ^{18}F into ^{18}O. The transition is labelled with β^+ endpoint energy and branching ratio, along with the branching ratio for the weak electron capture channel.

The remarkable clinical utility of ^{18}F is matched by its production scale: tens of thousands of curies are generated globally each day. This production relies on well-characterized nuclear data and optimized target engineering, making it an exemplary case of nuclear data enabling widespread and dependable medical isotope supply.

6.4.1 Reaction: $^{18}O(p,n)$

Figure 6.12 presents the excitation function for the $^{18}O(p,n)^{18}F$ reaction, compiled from multiple experimental datasets and consolidated in the IAEA-evaluated curve. This reaction is among the most extensively studied in nuclear medicine and serves as a reference case for reliable cross section evaluation. The excitation function displays a resonance-like profile: a sharp rise just above threshold, a resonance around 6 MeV with cross sections approaching 700 mb, and a gradual decline extending to approximately 20 MeV.

The well-defined shape of the excitation function enables precise optimization of beam energy for efficient ^{18}F production. Low-energy cyclotrons (e.g. 8–11 MeV) can still access the rising edge or the peak, while higher-energy systems (up to 30 MeV) can cover a broader section of the curve, provided a sufficiently thick target is used to degrade the beam through the energy range of interest. Alternatively, higher-energy machines may be tuned down or the target thickness adjusted to restrict energy deposition to the most favourable interval, improving material efficiency and reducing unwanted activation.

6.4.2 Target design and irradiation strategy

The evaluated excitation function provides a basis for tailoring irradiation parameters to maximize ^{18}F yield. In practice, the reaction is implemented using a water

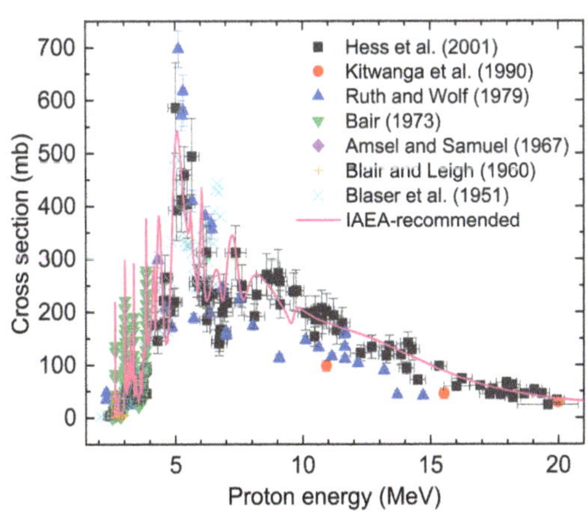

Figure 6.12. Excitation function for the $^{18}O(p,n)^{18}F$ reaction, compiled from multiple experimental studies [30–36], with the IAEA-evaluated curve overlaid. The cross section exhibits a resonance-like structure between 3 and 10 MeV, reaching peaks near 700 mb before gradually decreasing beyond 10 MeV. This profile supports efficient production using low- to mid-energy cyclotrons, typically operating below 20 MeV.

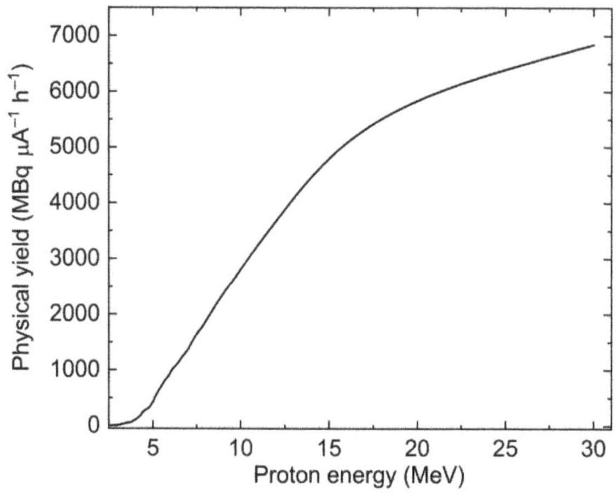

Figure 6.13. Physical yield for the $^{18}O(p,n)^{18}F$ reaction, based on IAEA-recommended cross section data in figure 6.12.

target chamber filled with ^{18}O-enriched water (typically >95% enrichment). Protons from a medical cyclotron, typically accelerated to 10–20 MeV, lose energy as they traverse the water, allowing the beam to span the most productive portion of the excitation curve. For example, a 17 MeV incident beam will decelerate through the 17–10 MeV range within the target, corresponding to the upper shoulder of the cross section where $\sigma_{(p,n)}$ remains just below 200 mb.

Figure 6.13 shows the corresponding physical yield calculated from the IAEA-evaluated data. The combination of a high cross section and the ability to sustain intense proton beams in liquid targets enables high production rates. Typical systems deliver 20–100 μA, and the thermal conductivity of water facilitates efficient heat dissipation at these current levels. In routine production, 1–3 mL of ^{18}O-enriched water is irradiated in a target body made from silver or niobium, sealed with thin HAVAR or titanium windows. Under standard conditions, beam currents up to 65 μA can yield several hundred GBq of ^{18}F within 1–2 h [37], making this one of the most productive and widely adopted production routes in nuclear medicine. After irradiation, the [^{18}F]fluoride ion is typically isolated via ion-exchange chromatography and used as the precursor in radiopharmaceutical synthesis.

The success of this approach depends critically on the availability of high-quality nuclear data, which supports accurate beam energy planning, reliable yield estimation and efficient use of costly ^{18}O-enriched target material. Furthermore, the consistency of the thick-target yield across different facilities enables scalable, decentralized production networks—essential for meeting the logistical demands of PET.

6.4.3 Impurity channels

While the $^{18}O(p,n)^{18}F$ reaction dominates in enriched water targets, minor impurities can arise from residual ^{16}O via the $^{16}O(p,\alpha)^{13}N$ channel. The evaluated

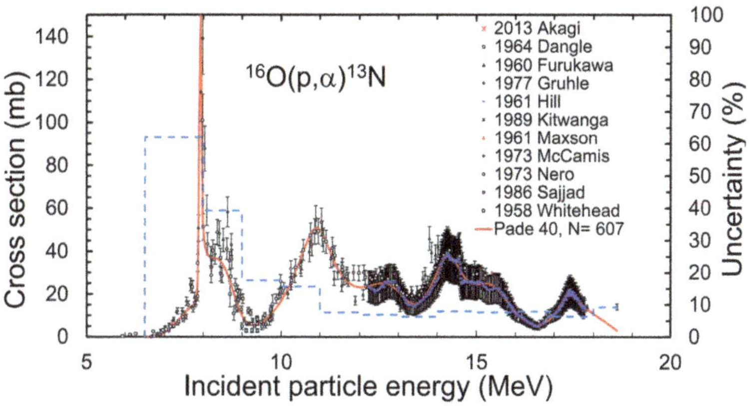

Figure 6.14. Evaluated excitation function (Padé fit) for the $^{16}O(p,\alpha)^{13}N$ reaction, shown alongside selected experimental data. The cross section exhibits a sharp peak of approximately 150 mb near 8 MeV and remains below 50 mb at higher energies. The estimated total uncertainty, including a 4% systematic component, is indicated by the dashed line (right-hand axis). Reprinted from [38] with permission from Elsevier.

excitation function for this reaction, shown in figure 6.14, exhibits a narrow resonance at 8 MeV with a peak cross section of 150 mb, followed by lower, fluctuating values below 50 mb up to 20 MeV. This profile means that ^{13}N production is most significant near 8 MeV and falls off at higher energies.

In production conditions, ^{13}N decays quickly ($t_{1/2}$ = 9.965 min) and is rarely present in significant quantities at the time of patient administration. However, its chemical behaviour can complicate radiochemical processing. Formed as nitrates and nitrites, ^{13}N species may strongly bind to the anion exchange resins used to trap [^{18}F]fluoride, particularly in the production of Na[^{18}F]F. These impurities are not easily removed by standard elution protocols and may require specific attention depending on the radiopharmaceutical formulation.

Fortunately, the cross section behaviour allows for mitigation: by tailoring the beam energy to maximize the $^{18}O(p,n)$ yield while avoiding the 8 MeV peak in the $^{16}O(p,\alpha)$ channel, impurity production can be minimized. Accurate nuclear data for both reactions therefore remain essential for balancing yield and purity in routine ^{18}F production.

6.4.4 Decay data and imaging

Much of the clinical utility of ^{18}F arises not only from its favourable decay characteristics but also from the high precision with which its decay parameters have been determined. The half-life, positron branching ratio and annihilation photon yield have all been measured with excellent accuracy and are compiled in evaluated decay datasets, such as those available from the NNDC.

In practical terms, these data support a wide range of operational and clinical requirements. The well-characterized half-life (109.77 min) enables accurate decay correction in activity measurements and supports precise scheduling for radiopharmaceutical synthesis, quality control and patient administration. The positron emission probability and annihilation photon yield are critical for PET scanner calibration and for ensuring consistency in image quantification across different clinical sites and instrumentation.

Because ^{18}F emits no significant prompt gamma-rays aside from the 511 keV annihilation photons, its dosimetry is straightforward, and the simplicity of its decay scheme—combined with precisely evaluated nuclear data—supports accurate dose calculation, scanner calibration and consistent imaging standards. These qualities facilitate regulatory approval, clinical adoption and international harmonization.

The case of ^{18}F illustrates how comprehensive and well-evaluated nuclear data enable efficient production, accurate dosimetry and reliable clinical implementation. Its success highlights what is possible when decay and reaction data are both precise and complete. However, many emerging medical isotopes still suffer from significant gaps or uncertainties in their nuclear data. The following section explores such cases, demonstrating the importance of continued efforts in decay data measurement and evaluation.

6.5 Decay data for medical isotopes

Understanding the decay characteristics of medical radioisotopes is just as important as mastering their production. Half-life and gamma-ray intensities are key nuclear data parameters. High-precision and accurate decay data are essential for dosimetry, activity calibration and ensuring safety in nuclear medicine. Recent research has therefore focused on refining the measured half-lives and decay schemes of emerging therapeutic and diagnostic isotopes. This section highlights two case studies—^{152}Tb and ^{225}Ac—to illustrate the importance of dependable decay data and ongoing efforts to improve it.

6.5.1 Case study 3: ^{152}Tb decay characterization

6.5.1.1 Half-life measurement

For ^{152}Tb, a positron-emitting isotope explored for PET imaging, an accurately determined half-life is crucial for correct activity quantification and patient-specific dose calculation. Among the terbium isotopes proposed for theranostic use, ^{152}Tb plays a prominent diagnostic role due to its β^+ decay to stable ^{152}Gd. It is one of four isotopes in a matched terbium quartet offering both diagnostic and therapeutic options. For ^{152}Tb to be reliably applied in clinical settings, particularly for PET imaging, its decay characteristics, most notably its half-life, must be determined with high accuracy and precision. Even small uncertainties in these parameters can affect activity quantification, image quality and dosimetry.

For decades, the half-life of ^{152}Tb was accepted from two measurements dating back to the 1960s, with considerable uncertainty. In an effort to address this, a recent campaign at ISOLDE remeasured the half-life using contemporary instrumentation and methodology [39]. The facility provides access to high-purity beams of exotic radionuclides through on-line mass separation. This capability is critical for isolating single isotopes such as ^{152}Tb from co-produced contaminants, thereby enabling decay studies on material with exceptional purity.

In this study, two independently prepared ^{152}Tb sources were generated via mass separation and subsequent radiochemical purification. Their decay was monitored using three complementary techniques: HPGe gamma-ray spectrometry, current-mode ionization chamber readings and liquid scintillation counting. Each technique offers distinct sensitivity and systematic behaviour, providing a robust cross-validation of results. Four half-life measurements in total were performed, two from HPGe detectors (applied to both sources) and one each from the ion chamber and the scintillation method.

The weighted average of these four independent results yielded a new half-life value of 17.8784(95) hours, significantly more precise than the older literature values. This new value is approximately 2.2% longer than the previously evaluated half-life and the tenfold improvement in uncertainty (to about 0.5%) reflects the benefits of modern measurement techniques and source purity. The discrepancy between old and new values corresponds to a ζ-score of approximately 3.8. This metric compares two values in terms of their combined uncertainties and a ζ-score above 3 typically indicates statistically significant disagreement. In this case, the result strongly suggests that the previously evaluated half-life underestimated the true value.

Figure 6.15 places the new measurements in context, showing the earlier reported values and highlighting the improved precision achieved. The close clustering of the four new values underpins the robustness of the result.

An additional outcome of this work was the confirmation of the high radio-nuclidic purity of the ^{152}Tb samples, thanks to the combination of ISOLDE's mass separation and radiochemical purification. This purity is essential for ensuring that the measured decay reflects only the target isotope, without interference from

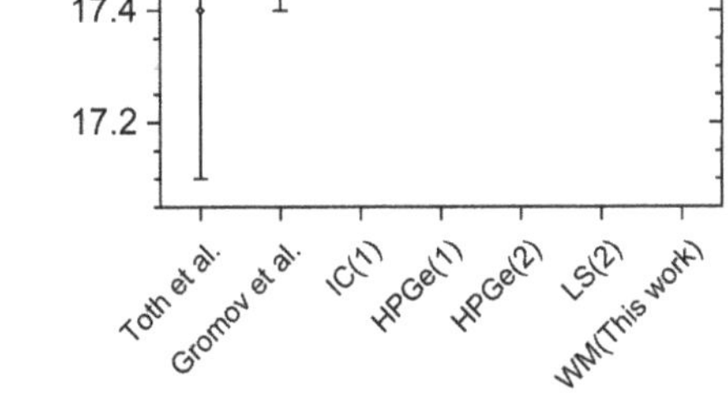

Figure 6.15. Summary of half-life determinations for ^{152}Tb. Earlier literature values from the 1960s are shown with their associated uncertainties, alongside four new measurements from 2023 using HPGe γ-ray spectrometry, ionization chamber (IC) and liquid scintillation (LS) counting. 'WM' denotes the weighted mean across all techniques. Numbers in brackets indicate the sample used for each measurement. (Reproduced from [39]. CC-BY 4.0.)

longer-lived contaminants. With this refined and reliable decay data, ^{152}Tb is better suited for use in PET imaging and as a calibration standard in nuclear medicine.

6.5.1.2 Coincidence spectroscopy

Half-life is only one facet of nuclear decay data. For a comprehensive understanding of an isotope's behaviour, it is equally important to know the decay scheme: the sequence of excited states populated in the daughter nucleus and the energies and intensities of all emissions—gamma-rays, beta particles and conversion electrons. These parameters affect both how the isotope is detected and how its radiation dose is deposited in tissue. In the case of ^{152}Tb, the decay populates a network of excited levels in ^{152}Gd, resulting in the emission of multiple gamma-rays and conversion electrons. The completeness of the decay scheme is essential for calculating the average energy per decay and understanding how different radiation components contribute to dosimetry.

Gamma–gamma spectroscopy

Two complementary experiments have recently refined the decay scheme of ^{152}Tb. The first used gamma–gamma coincidence spectroscopy to establish transitions between excited states in ^{152}Gd [40]. A high-purity ^{152}Tb source, produced at ISOLDE, was studied at the Institut Laue-Langevin (ILL) using an array of HPGe clover detectors. This enabled identification of cascades between levels, including weak and high-energy transitions, that were previously uncertain or missing due to the pandemonium effect.

Figure 6.16 shows a partial decay scheme illustrating representative transitions established through the gamma–gamma study. Prior to this work, the most recent full evaluation, based on studies culminating in 2003 [41], suffered from incomplete coverage of weak or high-energy branches—a classic manifestation of the pandemonium effect, where undetected transitions from high-lying levels result in overestimated feeding to low-lying states. The coincidence data now confirm or add numerous transitions and clarify the population of high-energy levels, improving the mapping of the beta strength function.

Electron–gamma spectroscopy

The second study [42] employed electron–gamma coincidence spectroscopy using the LOHENGRIN spectrometer. This approach allows detection of conversion electrons emitted in nuclear de-excitations, providing information on transitions that lack a gamma-ray branch, such as electric monopole E0 transitions (see section 6.6 of [43]). These transitions cannot be observed with gamma-ray spectroscopy alone, but they play a key role in completing the decay scheme by accounting for de-excitation pathways that involve only electron emission. By quantifying the intensity of such transitions, the population and depopulation of excited nuclear levels can be more effectively traced, which is essential for inferring the beta strength function. This, in turn, improves understanding of the energy released per decay and enhances the accuracy of dose calculations in medical applications.

Figure 6.17 offers a clear side-by-side comparison of electron and gamma detection, underscoring the complementary nature of these techniques. It shows how E0 transitions, such as those at 433 keV and 615 keV, are undetectable in

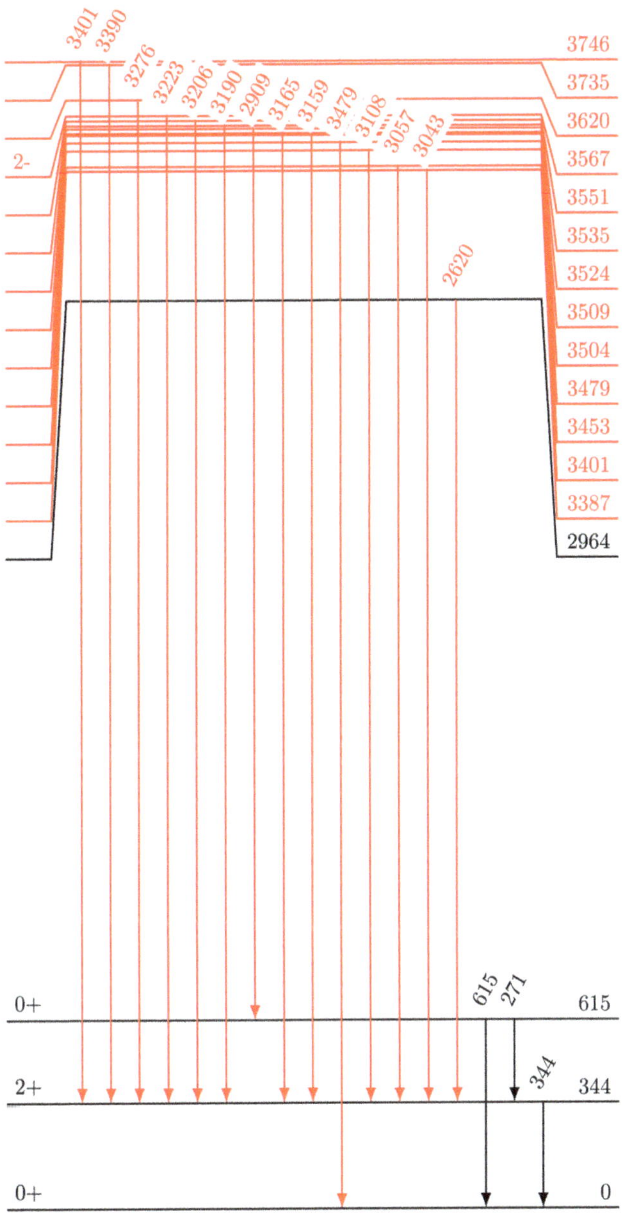

Figure 6.16. Partial level scheme for the decay of ^{152}Tb into ^{152}Gd, highlighting key transitions identified in the spectroscopy campaign. Levels (energies in keV) in the daughter ^{152}Gd nucleus are shown as horizontal lines, with arrows indicating gamma transitions between them. Each arrow is labelled with the transition energy. Levels and transitions shown in red were newly identified in this study and are not included in existing evaluated data. (Reproduced from [40]. CC-BY 4.0.)

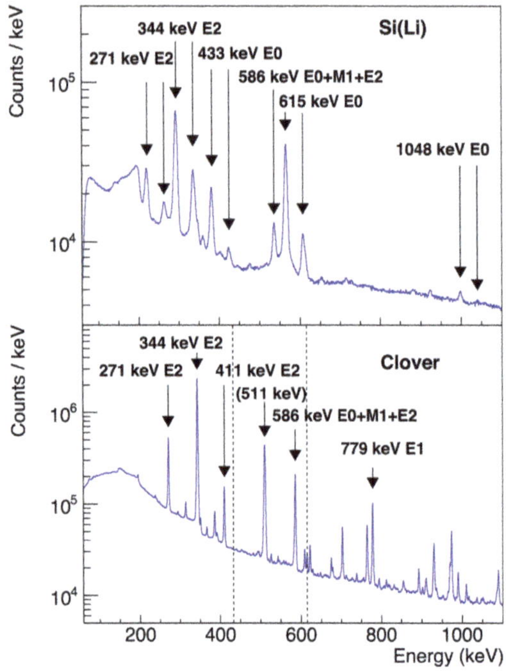

Figure 6.17. Singles spectra recorded in the electron–gamma decay spectroscopy of ^{152}Tb. The top panel shows conversion electron signals detected by a Si(Li) detector, highlighting distinct K- and L-shell peaks from transitions such as 344, 433, 586 and 615 keV. Notably, the electric monopole E0 transitions at 433 keV and 615 keV, which have no gamma-ray branch, are clearly visible only in the electron spectrum. The bottom panel displays the corresponding HPGe gamma-ray spectrum, where no peaks are observed at the E0 transition energies (dashed lines), illustrating their invisibility to gamma-only spectroscopy. (Reproduced from [42]. CC-BY 4.0.)

gamma-ray spectra but are clearly observed in the electron spectra. This observation confirms the added value of electron–gamma spectroscopy in capturing transitions that would otherwise remain hidden, supporting the construction of a more complete decay scheme and enabling more reliable modelling of the beta strength function.

In addition, measurements of internal conversion coefficients provided valuable insights into the character of several transitions. These coefficients, which quantify the probability of internal conversion relative to gamma emission, are sensitive to the multipolarity of the transition. By comparing experimental values with theoretical predictions for different multipolarities, it was possible to clarify or assign multipolarities for transitions that were previously uncertain. This step is essential for constructing a physically consistent decay scheme and for improving models of energy release per decay—particularly in cases where highly converted or low-intensity transitions contribute significantly to dose calculations in medical applications.

The combination of these two studies enables more accurate determination of the average energy released per decay and how it is partitioned among beta particles, gamma-rays and conversion electrons. Such detail is critical in medical contexts, where reliable nuclear data underpin patient-specific dosimetry and radiation safety. These recent studies exemplify how modern spectroscopy—both gamma–gamma and electron–gamma—resolves long-standing ambiguities in decay schemes and strengthens the nuclear data foundation for clinical use of isotopes such as ^{152}Tb.

6.5.2 Case study 4: ^{225}Ac decay chain

The decay of ^{152}Tb is essentially a one-step process that leads directly to a stable daughter, but some therapeutically important radionuclides follow far more elaborate decay routes. A textbook case is ^{225}Ac, an α-emitter that lies at the heart of several targeted alpha therapy (TAT) protocols. With a half-life of roughly ten days, ^{225}Ac is long-lived enough for production, shipping and labelling, yet short enough to deliver a clinically useful dose. Importantly, it does not decay to stability in a single step; instead it passes through a cascade of short-lived daughters. The chain terminates at quasi-stable ^{209}Bi, but along the way at least eight intermediate nuclides contribute additional radiations that complicate dosimetry and imaging.

Figure 6.18 shows the cascade from ^{225}Ac to quasi-stable ^{209}Bi. In the dominant path, four α decays are threaded by two β^- decays. Two nodes in the chain need special attention:

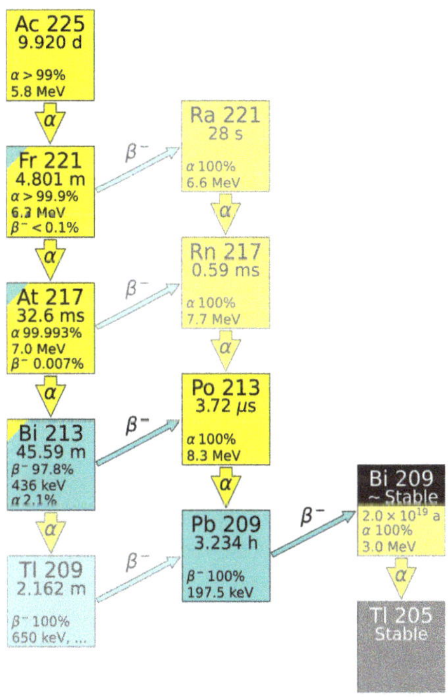

Figure 6.18. Decay chain of ^{225}Ac showing alpha and beta transitions. Each nuclide is represented by a box labelled with its name, half-life and principal decay modes. The fill colour indicates the dominant decay type: yellow for alpha emission, blue-green for beta-minus decay and black for stable nuclei. Minority decay branches are marked by coloured triangles inset at the upper-left corner of the box. Arrows indicate the direction and type of each transition. Faded boxes and arrows correspond to less prominent branches of the decay chain.

- **^{221}Fr** $\left(T_{1/2} = 4.80 \text{ min}\right)$ α-decays with >99.9 % probability but carries a tiny β^- branch (<0.1 %). Its 218.0(1) keV gamma ($I_\gamma \approx 11.44(12)$ %) is the most intense single-photon signature in the early part of the chain, yet its quoted intensity still carries a few-percent uncertainty, complicating high-accuracy activity assays.
- **^{213}Bi** ($T_{1/2} = 45.59$ min) provides the main β^- link (97.8 %$\rightarrow$^{213}Po) but also an α branch (2.1 % $\rightarrow^{209}$ Tl). The latter adds extra steps and dose-relevant emissions from ^{209}Tl and ^{209}Pb. The 440.45(1) keV gamma transition from ^{213}Bi ($I_\gamma \approx 25.9(2)$ %) is the brightest photon in the whole sequence, yet its quoted intensity likewise needs refinement for metrological work.

From a clinical standpoint the cascade is double-edged. Multiple high linear energy transfer (LET) α emissions amplify tumour cell kill, but the recoil from each α step can physically displace daughter atoms from the chelator or nanoparticle carrier, redistributing activity to healthy tissue. Strategies such as nanoparticle encapsulation or local administration are under active investigation, but all rely on detailed, quantitative decay data.

6.5.2.1 Half-life measurement

Historically, ^{225}Ac was assigned a rounded half-life of 10 days, derived from two early studies with no formal uncertainty budget. Recognizing the need for a metrologically sound value, Pommé e*t al* [44] conducted a six-method campaign that included: α-particle counting with a planar silicon detector at a defined small solid angle (αDSA) and in a nearly-2π geometry (2$\pi\alpha$); 4$\pi\alpha\beta$ counting using a windowless CsI sandwich spectrometer (CsI) and a large pressurized proportional counter (LPPC); and γ-ray spectrometry performed with both an HPGe detector (HPGe) and a NaI(Tl) well detector (NaI). Each technique tracked the decay over at least six half-lives, with rigorous corrections for dead time, pile-up, background and efficiency. All decay curves agreed within 0.1%, yielding a weighted mean half-life of 9.920(3) days for ^{225}Ac.

Figure 6.19 visualizes the agreement. The revision—0.8% lower than the long-standing 10 d estimate—feeds directly into calibrations, logistics and patient dosimetry. A similar level of rigour will ultimately be required for all daughters.

6.5.2.2 Decay chain spectroscopy

While the parent half-life is now well established at 9.920(3) d, uncertainties persist in the spectroscopic properties of key short-lived descendants such as ^{213}Bi and ^{221}Fr. For example, early evaluations reported the alpha-particle energies of ^{213}Bi with uncertainties of up to 10 keV and branching ratios lacking quantified uncertainties [45]. To improve this, Marouli *et al* [46] performed a dedicated alpha-spectrometry study using recoil atoms from an open ^{225}Ac source. Their work led to refined energies of $E_{\alpha,0} = 5878(4)$ keV and $E_{\alpha,1} = 5560(4)$ keV and established relative emission probabilities of $P_{\alpha,0} = 91.55(11)\%$ and $P_{\alpha,1} = 8.45(11)\%$, with a branching factor of $P_\alpha = 2.140(10)\%$. A comparison of literature and measured peak energies is shown in figure 6.20, highlighting the improved consistency across the ^{225}Ac decay chain.

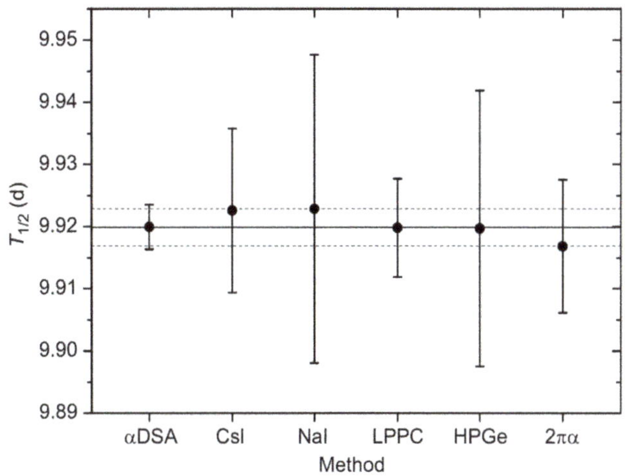

Figure 6.19. Independent half-life determinations for ^{225}Ac. The solid line shows the weighted mean 9.920(3) d. The dashed lines indicate the standard uncertainty ($k = 1$), corresponding to one standard deviation. (Reproduced with permission from [44]. Copyright 2012 Elsevier.)

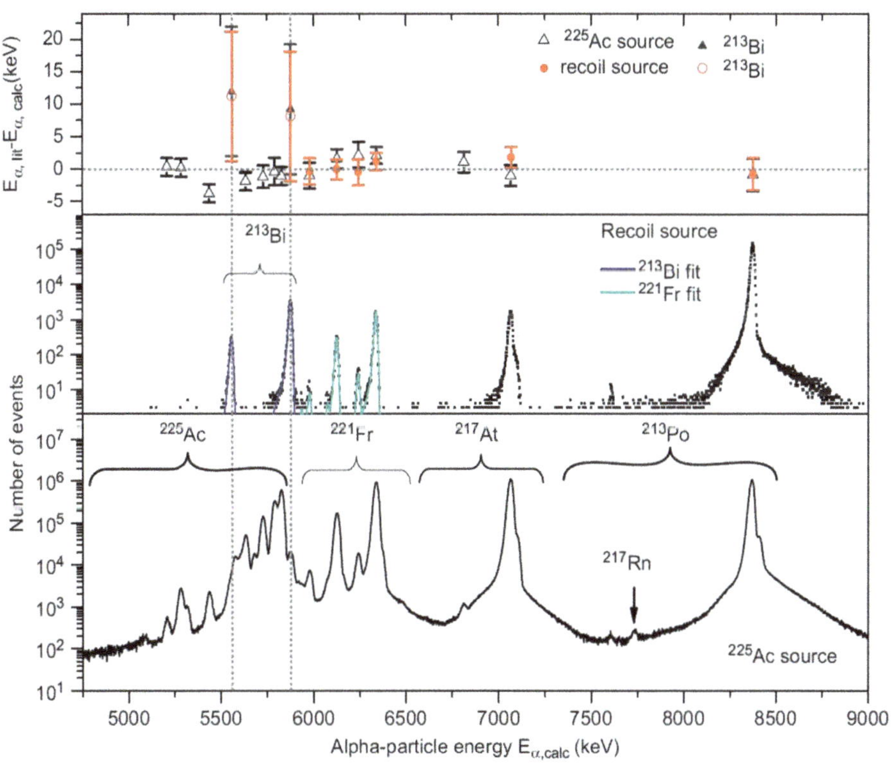

Figure 6.20. Alpha spectra from an ^{225}Ac source (bottom panel) and from collected recoil atoms (middle panel), highlighting the dominant ^{213}Bi peaks. The top panel shows deviations of measured energies from previously published values. (Reproduced with permission from [46]. Copyright 2013 Elsevier.)

Despite this progress, significant gaps remain. Gamma-ray transitions from ^{221}Fr and ^{213}Bi—particularly weak lines or those with ambiguous multipolarities—are still insufficiently resolved. Moreover, existing measurements are typically limited to alpha spectroscopy, without coincident gamma detection. To comprehensively address these issues, a modular decay spectroscopy campaign has been initiated at the ISOLDE Decay Station (IDS) [47], aiming to resolve unresolved multipolarities, refine level schemes and improve decay heat and dose modelling through high-statistics coincidence data.

The IDS set-up enables coincident charged-particle and gamma-ray detection using a modular configuration. Beams from ISOLDE (see section 6.2.1) are implanted onto aluminised Mylar tape, allowing for rapid source transport and removal of long-lived activity, which facilitates high-purity, time-resolved measurements. For high-resolution studies of alpha-emitting decay chains such as that of ^{225}Ac, coincidence spectroscopy techniques are typically employed to disentangle complex level schemes and quantify weak or highly converted transitions. A common experimental set-up combines charged-particle detectors—such as a plastic scintillator and an annular silicon detector—with a high-efficiency gamma-ray detection array, often comprising HPGe clover detectors. This configuration enables simultaneous identification of α, β and γ emissions, while the adjustable geometry of the clover array supports angular correlation measurements for multipolarity assignments.

To investigate short-lived nuclides within a decay chain, such as ^{221}Fr or ^{213}Bi, isotopically pure beams may be delivered directly to the detection station for prompt spectroscopy. For longer-lived parents such as ^{225}Ac, activity is typically accumulated on a transportable tape system and monitored over extended time intervals to observe the in-growth and equilibrium behaviour of daughter products. Using a consistent detector arrangement across all measurements allows for internally self-consistent comparisons of branching ratios, level populations and internal conversion coefficients.

Particular attention is often given to transitions of diagnostic or dosimetric relevance. The 218 keV line from ^{221}Fr and the 440 keV line from ^{213}Bi, for example, are routinely used in radiopharmaceutical assays and dose calibrations, yet still carry non-negligible uncertainties in their intensities. Weakly populated or highly converted transitions, such as the E2–E2 cascade at 465–1567 keV in ^{209}Tl, can also be resolved through high-statistics coincidence data and angular correlation analysis. These measurements contribute directly to improved decay heat modelling, radioprotection planning and generator calibration. In the context of TAT, refined decay data also support the development of molecular containment strategies, such as nanoparticle encapsulation or ligand systems designed to retain recoiling daughter nuclei.

6.6 Translating nuclear data to medical isotopes

Every diagnostic scan or therapeutic dose that reaches the clinic rests on an unbroken chain of nuclear data. The chain begins in the laboratory, where cross sections and decay properties are measured and ends in a hospital hot-cell or dose calibrator. Between those points lies a self-correcting information loop that turns scattered experiments into the reliable numbers producers and clinicians need.

6.6.1 Compilation of measurements

Freshly published cross sections flow into the IAEA's EXFOR database, while new half-life or gamma intensity determinations are logged by projects such as the Decay Data Evaluation Project or the NNDC These repositories preserve the original experimental detail, including quoted uncertainties.

6.6.2 Evaluation and modelling

Specialist evaluator teams compare overlapping datasets, correct for known biases and fit nuclear reaction codes (e.g. TALYS, EMPIRE) where measurements are sparse. The outcome is a set of *recommended* values with traceable uncertainties, published in master libraries such as ENDF, JEFF and Evaluated Photonuclear Data Library. For medical work, the IAEA compiles an application-specific Medical Charged-Particle Cross Section Database, while model-based libraries like TENDL supply provisional numbers when experiments are lacking.

6.6.3 Dissemination and use

Evaluated nuclear data libraries are processed by tools such as NJOY, which convert them into formats suitable for Monte Carlo transport codes including MCNP, GEANT4, FLUKA and PHITS. These codes are widely used to design cyclotron targets, predict thick-target yields and assess shielding requirements. Decay data flow into calibration software and online platforms such as NuDat and the IAEA LiveChart, supporting dose calibrator settings and patient dosimetry. Whether estimating ^{67}Cu yields from ^{70}Zn(p,α) reaction or checking the gamma emission probability of the 218 keV line in ^{221}Fr, users depend, often unknowingly, on these evaluated datasets.

6.6.4 Feedback and refinement

When evaluations highlight gaps or contradictions, such as the spread in ^{68}Zn(γ,p)^{67} Cu data, new experiments are commissioned. The results loop back into EXFOR, the cycle turns again and recommended values tighten. The process is continuous: as emerging radionuclides such as ^{161}Tb and ^{225}Ac move toward clinical use, each step of the data chain is revisited and refined.

Accurate cross sections maximize yield and minimize impurities; precise decay data underpin dose calibrations and patient safety. The sustained, international effort to measure, evaluate and publish nuclear data are therefore a hidden, but essential, pillar of modern nuclear medicine. Continued collaboration between experimentalists, evaluators and end-users will ensure that tomorrow's radiopharmaceuticals are built on numbers every bit as dependable as today's.

References

[1] Ivashkin A, Abdurashitov D, Baranov A, Guber F, Morozov S, Musin S, Strizhak A and Tkachev I 2023 Testing entanglement of annihilation photons *Sci. Rep.* **13** 7559

[2] Parashari S, Bosnar D, Friščić I, Kožuljević A M, Kuncic Z, Žugec P and Makek M 2024 Closing the door on the “puzzle of decoherence” of annihilation quanta *Phys. Lett.* B **852** 138628

[3] Bordes J, Brown J R, Watts D P, Bashkanov M, Gibson K, Newton R and Zachariou N 2024 First detailed study of the quantum decoherence of entangled gamma photons *Phys. Rev. Lett.* **133** 132502

[4] Moskal P *et al* 2025 Non-maximal entanglement of photons from positron–electron annihilation demonstrated using a novel plastic PET scanner *Sci. Adv.* **11** eads3046

[5] Moskal P *et al* 2024 Positronium image of the human brain *in vivo Sci. Adv.* **10** eadp2840

[6] Boschi A, Uccelli L and Martini P 2019 A picture of modern Tc-99m radiopharmaceuticals: production, chemistry and applications in molecular imaging *Appl. Sci.* **9** 2526

[7] Caobelli F, Wollenweber T, Bavendiek U, Kühn C, Schütze C, Geworski L, Thackeray J T, Bauersachs J, Haverich A and Bengel F M 2016 Simultaneous dual-isotope solid-state detector SPECT for improved tracking of white blood cells in suspected endocarditis *Eur. Heart J.* **38** 436–43

[8] Audenhaege K V, Holen R V, Vandenberghe S, Vanhove S, Metzler S D and Moore S C 2015 Review of SPECT collimator selection, optimization, and fabrication for clinical and preclinical imaging *Med. Phys.* **42** 4796–813

[9] Tashima H and Yamaya T 2022 Compton imaging for medical applications *Radiol. Phys. Technol.* **15** 187–205

[10] Danieli R, Stella M, Leube J, Tran-Gia J, Marin C, Uribe C F, Vanderlinden B, Reynaert N, Flamen P and Levillain H 2023 Quantitative ^{177}Lu SPECT/CT imaging for personalized dosimetry using a ring-shaped CZT-based camera *Eur. J. Nucl. Med. Mol. Imaging Phys.* **10** 64

[11] Lhommel R, Elmbt L V, Goffette P, den Eynde M V, Jamar F, Pauwels S and Walrand W 2010 Feasibility of ^{90}Y TOF PET-based dosimetry in liver metastasis therapy using SIR-spheres *Eur. J. Nucl. Med. Mol. Imaging* **37** 1654–62

[12] Elschot M, Vermolen B J, Lam M G E H, de Keizer B, van den Bosch M A A J and de Jong H W A M 2013 Quantitative comparison of PET and bremsstrahlung SPECT for imaging the *in vivo* yttrium-90 microsphere distribution after liver radioembolization *PLoS One* **8** 1–10

[13] Laughhunn L, Botkin C, Hubble W, Hewing D, Turner J, Muzaffar R, Robertson E and Osman M 2014 Xofigo: an exciting new therapy for prostate cancer *J. Nucl. Med.* **55** 2707

[14] Weineisen M *et al* 2015 ^{68}Ga- and ^{177}Lu-labeled PSMA I&T: optimization of a PSMA-targeted theranostic concept and first proof-of-concept human studies *J. Nucl. Med.* **56** 1169–76

[15] Banerjee S R *et al* 2020 Preclinical evaluation of $^{203/212}$Pb-labeled low-molecular-weight compounds for targeted radiopharmaceutical therapy of prostate cancer *J. Nucl. Med.* **61** 80–8

[16] Robertson A K H, Lobbezoo A, Moskven L, Schaffer P and Hoehr C 2019 Design of a thorium metal target for ^{225}Ac production at TRIUMF *Instruments* **3** 18

[17] Jenkins D 2020 *Radiation Detection for Nuclear Physics* (IOP Publishing)

[18] Hermanne A *et al* 2018 Reference cross sections for charged-particle monitor reactions *Nucl. Data Sheets* **148** 338–82

[19] Koning A, Hilaire S and Goriely S 2023 TALYS: modeling of nuclear reactions *Eur. Phys. J.* A **59** 131

[20] Tajima T and Dawson J M 1979 Laser electron accelerator *Phys. Rev. Lett.* **43** 267–70

[21] Mangles S P D *et al* 2004 Monoenergetic beams of relativistic electrons from intense laser–plasma interactions *Nature* **431** 535–8

[22] Wheldon C, Wheldon T K, Bishop J, Chiu Y -L, Phoenix B and Hampel D 2024 A Birmingham laboratory portrait *Nucl. Phys. News* **34** 4–8

[23] Kastleiner S, Coenen H H and Qaim S M 1999 Possibility of production of ^{67}Cu at a small-sized cyclotron via the (p,α)-reaction on enriched ^{70}Zn *Radiochim. Acta* **84** 107–10

[24] Levkovskij V N 1991 *Activation Cross Section Nuclides of Average Masses (*A *= 40–100) by Protons and Alpha-Particles with Average Energies (E = 10–50 MeV)* (INTER–VESTI)

[25] Sioufi A E, Erdös P and Stoll P 1957 γ, np)-Prozesse am Mo^{92} und Zn^{66} *Helv. Phys. Acta* **30** 264

[26] Koning A, Hilaire S and Goriely S 2023 TALYS-2.0 simulation of nuclear reactions *IAEA* https://nds.iaea.org/talys/

[27] Litvinenko V N *et al* 1997 Gamma-ray production in a storage ring free-electron laser *Phys. Rev. Lett.* **78** 4569–72

[28] Song J *et al* 2024 Photonuclear cross sections for ^{197}Au: an update on the gold standard *Phys. Rev.* C **110** 064604

[29] Thiep T D, An T T, Khai N T, Vinh N T, Cuong P V, Gangrski L P, Belov A G and Maslov O D 2012 Determination of the total bremsstrahlung photon flux from electron accelerators by simultaneous activation of two monitors *Phys. Part. Nucl. Lett.* **9** 648–55

[30] Hess E, Takács S, Scholten B, Tárkányi F, Coenen H H and Qaim S M 2001 Excitation function of the ^{18}O(p,n)^{18}F nuclear reaction from threshold up to 30 MeV *Radiochim. Acta* **89** 357–62

[31] Kitwanga S W, Leleux P, Lipnik P and Vanhorenbeeck J 1990 Production of $^{14,\,15}$O, ^{18}F, and ^{19}Ne radioactive nuclei from (p,n) reactions up to 30 MeV *Phys. Rev.* C **42** 748–52

[32] Ruth T J and Wolf A P 1979 Absolute cross sections for the production of ^{18}F via the ^{18}O(p, n)^{18}F reaction *Radiochim. Acta* **26** 21–4

[33] Bair J K 1973 Total neutron yields from the proton bombardment of $^{17,\,18}$O *Phys. Rev.* C **8** 120–3

[34] Amsel G and Samuel D 1967 Microanalysis of the stable isotopes of oxygen by means of nuclear reactions *Anal. Chem.* **39** 1689–98

[35] Blair J M and Leigh J J 1960 Total cross sections of the O^{18}(p,α)N^{15} and O^{18}(p,n)F^{18} reactions *Phys. Rev.* **118** 495–8

[36] Blaser J P, Boehm F, Marmier P and Scherrer P 1951 Excitation functions of the (p,n) reaction for light elements *Helv. Phys. Acta* **24** 465

[37] Blok S, Wängler C, Bartenstein P, Jurkschat K, Schirrmacher R and Lindner S 2023 Good practices for the automated production of ^{18}F-SiFA radiopharmaceuticals *EJNMMI Radiopharm. Chem.* **8** 25

[38] Hermanne A, Tárkányi F T, Ignatyuk A V, Takács S and Capote R 2021 Upgrade of IAEA recommended data of selected nuclear reactions for production of PET and SPECT isotopes *Nucl. Data Sheets* **173** 285–308

[39] Collins S M *et al* 2023 Determination of the terbium-152 half-life from mass-separated samples from CERN-ISOLDE and assessment of the radionuclide purity *Appl. Radiat. Isot.* **202** 111044

[40] O'Sullivan E B *et al* 2025 Towards complete decay spectroscopy of ^{152}Tb *Radiat. Phys. Chem.* **232** 112641

[41] Adam J *et al* 2003 Properties of ^{152}Gd collective states *Eur. Phys. J.* A **18** 605–26

[42] O'Sullivan E B *et al* 2025 Electron-gamma decay spectroscopy of ^{152}Tb *Phys. Scr.* **100** 065308

[43] Jenkins D G and Wood J L 2021 *Nuclear Data: A Primer* (IOP Publishing)

[44] Pommé S *et al* 2012 Measurement of the ^{225}Ac half-life *Appl. Radiat. Isot.* **70** 2608–14

[45] Huang X and Wang B 2007 Evaluation of decay data of ^{213}Bi *Nucl. Sci. Tech.* **18** 261–7

[46] Marouli M *et al* 2013 Decay data measurements on ^{213}Bi using recoil atoms *Appl. Radiat. Isot.* **74** 123–7

[47] Cocolios T E *et al* 2023 Detailed decay spectroscopy of ^{225}Ac and its daughters to support its use in medical applications, *Technical Report* CERN

Chapter 7

Epilogue

David Jenkins

In bringing this volume to a close, it is perhaps helpful to summarize what we have learned and provide some pointers to how to take your interest in nuclear data further. First, we have seen how nuclear data are a real world application of more fundamental science topics such as nuclear structure and nuclear reactions. Second, we have seen how diverse nuclear data can be from the point of view of how it underpins so many aspects of particularly energy generation and medical isotope production. Each of these applications brings substantial societal and economic benefit. Further, we have seen how many nuclear data needs frequently require high precision and high accuracy. Achieving this is challenging and drives parallel developments in accelerator technology, radiation detection technology, Monte Carlo simulation as well as nuclear theory.

Let us turn now to how to follow up in more depth on the various topics presented in this book. For those who want to read more on nuclear structure, we recommend the earlier *Nuclear Data* books from this IOP Book series [1–3]. For those who really want to get a deep understanding, in terms of a more fundamental grounding in the quantum mechanics related to nuclear structure, there are two other relevant books in the series [4, 5]. For the theory of nuclear reactions, there are quite a few textbooks available (many of which are sadly out-of-print at this stage) but the more recent, and perhaps more useful, are those aligned more towards nuclear astrophysics—the origin of the elements [6–8]. There are definite parallels between nuclear astrophysics and nuclear data in some of the approaches, e.g. the networks of reactions that synthesize nuclei in stars and explosive stellar environments with those networks of reactions which generate the minor actinides in reactors.

For specific applications of nuclear data, the main resource is the scientific literature as there are not really any textbooks in this area, other than the present work. Accordingly, we have provided a significant number of references throughout this book as a jumping off point. Other ports of call are technical reports from organizations such as the IAEA and NEA. A further resource would be international conferences and workshops. The premier example is the Nuclear Data for

doi:10.1088/978-0-7503-5102-7ch7

© IOP Publishing Ltd 2025. All rights, including for text and data mining (TDM), artificial intelligence (AI) training, and similar technologies, are reserved.

Science and Technology Conference which is held every three years; it was last held in Madrid in June 2025.

As detailed throughout the book, there are many nuclear data resources and databases freely accessible. This allows someone to readily explore ideas and cement knowledge gained here through experimentation.

Wherever your nuclear data journey takes you, we wish you a happy and fulfilling career.

References

[1] Jenkins D G and Wood J L 2021 *Nuclear Data: A Primer* (IOP Publishing)
[2] Jenkins D G and Wood J L 2023 *Nuclear Data: A Collective Motion View* (IOP Publishing)
[3] Jenkins D G and Wood J L 2024 *Nuclear Data: An Independent-Particle Motion View* (IOP Publishing)
[4] Heyde K and Wood J L 2019 *Quantum Mechanics for Nuclear Structure: A Primer* (IOP Publishing)
[5] Heyde K and Wood J L 2020 *Quantum Mechanics for Nuclear Structure: An Intermediate Level View* (IOP Publishing)
[6] Rolfs C and Rodney W S 1988 *Cauldrons in the Cosmos* (University of Chicago Press)
[7] Iliadis C 2007 *Nuclear Physics of Stars* (Wiley)
[8] Thompson I J and Nunes F M 2012 *Nuclear Reactions for Astrophysics* (Cambridge University Press)

Appendix A

Stopping powers

Stopping power describes the energy loss per unit length of charged particles travelling through matter, primarily due to interactions with electrons (electronic stopping) and atomic nuclei (nuclear stopping). Expressed as $S(E) = -\mathrm{d}E/\mathrm{d}x$, the stopping power can be represented compactly by incorporating critical corrections to the Bethe–Bloch formula (excluding charge-state corrections) [1]:

$$S(E) = \frac{\kappa Z_2}{\beta^2} Z_1^2 L(\beta), \tag{A.1}$$

where Z_1 and Z_2 represent atomic numbers of incident and target atoms, $\beta = v/c$ is the particle velocity relative to the speed of light, $\kappa = 4\pi r_0^2 m_e c^2$ involves the classical electron radius $r_0 = e^2/(m_e c^2)$, electron charge e and electron mass m_e. The stopping number $L(\beta)$ incorporates various energy loss mechanisms:

$$L(\beta) = L_0(\beta) + Z_1 L_1(\beta) + Z_1^2 L_2(\beta) + \cdots \tag{A.2}$$

where the term $L_0(\beta)$ includes the fundamental energy loss contributions described by the Bethe–Bloch theory. The term $L_1(\beta)$, known as the Barkas or Z_1^3 correction [2], accounts for the dependence of stopping power on the sign of the particle charge. The term $L_2(\beta)$, referred to as the Bloch or Z_1^4 correction [3], accounts for higher-order corrections to the Bethe formula, ensuring consistency with Bohr's classical result for slow heavy projectiles.

The Bethe–Bloch equation provides the classical description of the stopping power due to inelastic collisions with atomic electrons and is given by

$$S(E) = \frac{4\pi e^4 Z_2}{m_e v^2} Z_1^2 \left[\ln \frac{2 m_e v^2}{\langle I \rangle} - \ln(1 - \beta^2) - \beta^2 \right], \tag{A.3}$$

where $\langle I \rangle$ is the averaged excitation potential per electron, which can be approximated empirically. For example, following [4], it can be expressed as $\langle I \rangle = (9.76 + 58.8 Z^{-1.19}) Z$ eV. While this stopping power equation provides accurate predictions for high-energy

doi:10.1088/978-0-7503-5102-7ch8 © IOP Publishing Ltd 2025. All rights, including for text and data mining (TDM), artificial intelligence (AI) training, and similar technologies, are reserved.

charged particles, it does not account for relativistic effects, atomic shell corrections, or medium polarization.

Fano extended [5] the Bethe–Bloch equation by introducing additional terms to account for atomic shell effects at low energies and polarization corrections at high energies. The modified equation includes the shell correction term C/Z_2, which accounts for atomic structure effects and the density correction term $\delta/2$, which considers polarization effects in dense media:

$$S(E) = \frac{4\pi e^4 Z_2}{m_e v^2} Z_1^2 \left[\ln \frac{2m_e v^2}{\langle I \rangle} - \ln(1 - \beta^2) - \beta^2 - \frac{C}{Z_2} - \frac{\delta}{2} \right]. \tag{A.4}$$

Introducing a relativistic function simplifies the above equation:

$$f(\beta) \equiv \ln \left[2m_e c^2 \frac{\beta^2}{(1 - \beta^2)} \right] - \beta^2, \tag{A.5}$$

and yields the stopping power as

$$S(E) = \frac{4\pi r_0^2 m_e c^2 Z_2}{\beta^2} Z_1^2 \left[f(\beta) - \ln\langle I \rangle - \frac{C}{Z_2} - \frac{\delta}{2} \right], \tag{A.6}$$

which can be simplified into the compact form of equation (A.1).

A.1 Charged particle range

Particle range in a medium, the distance a particle travels until it loses its entire kinetic energy E_0, is computed by integrating the reciprocal of the stopping power:

$$R = \int_0^{E_0} \frac{\mathrm{d}E}{S(E)}. \tag{A.7}$$

For high-energy charged particles, the range can be estimated using the Bethe–Bloch equation, while at lower energies, empirical corrections are often necessary. The range is an important parameter in experimental nuclear physics, particularly in cross section measurement experiments where precise knowledge of the energy loss in the target is required.

A.2 Computational methods for stopping power, range and straggling

Predicting charged-particle energy loss and penetration depth relies on a combination of theoretical models, empirical formulas and Monte Carlo or deterministic simulation tools. While the Bethe–Bloch equation and its extensions remain fundamental, practical applications often employ software such as SRIM and GEANT4 to account for material properties, nuclear stopping and statistical effects. As an illustrative example using SRIM [6], the total stopping power may be expressed as

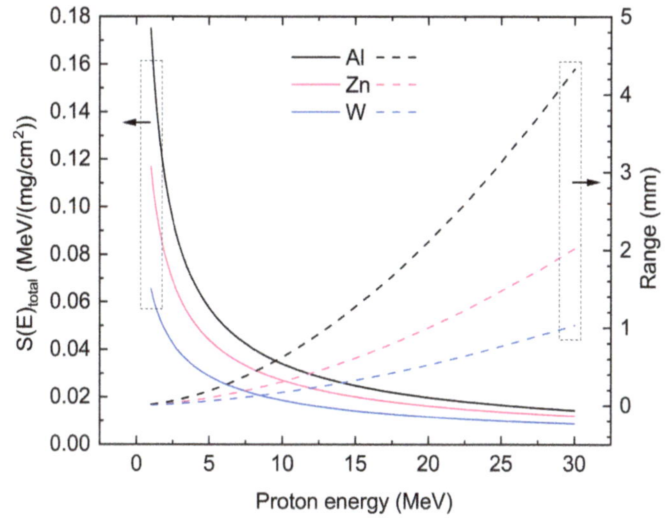

Figure A.1. Total stopping power of selected materials for protons, $S(E)_{\text{total}}$, (solid lines, left axis) and the range of protons in these materials (dashed lines, right axis) over a representative energy range, calculated by SRIM. Boxes and arrows highlight the correspondence between the two datasets and their respective y-axes.

$$S(E)_{\text{total}} = S(E)_{\text{elec}} + S(E)_{\text{nucl}}, \tag{A.8}$$

where $S(E)_{\text{elec}}$ follows the Bethe–Bloch formalism (see equation (A.1)) and $S(E)_{\text{nucl}}$ represents energy loss due to elastic collisions with nuclei—a contribution that grows significant at lower energies. Figure A.1 illustrates typical outputs of stopping power and particle range across materials, a form of analysis similarly performed within various transport codes. Such results are essential for designing experimental set-ups, targets and beam-energy profiles in precision nuclear measurements.

The statistical spread of energy loss—known as straggling—is also modelled in these tools. The electronic straggling at high particle energies can be estimated using the relativistic Bohr formula [7]:

$$Q = 4\pi e^4 Z_1^2 Z_2 N \Delta x \frac{(1 - \beta^2/2)}{(1 - \beta^2)}, \tag{A.9}$$

where N is the atomic density of the target and Δx is the distance travelled by the projectile. At lower velocities, Firsov's prediction [8] is used, which states that below the Fermi velocity (approximately 25keV amu^{-1}), straggling is proportional to the projectile energy until it intersects with the Bohr straggling value, at which point it follows the Bohr limit.

Additionally, simulations often do not automatically incorporate crystalline effects such as channelling—when ions traverse through lattice channels—requiring specialized modelling to capture reduced or enhanced straggling due to ordered atomic alignment or dechannelling phenomena. For further discussion of stopping power, particle interactions and related modelling approaches, see [9].

In summary, computational frameworks ranging from analytical formulas to comprehensive simulation environments, e.g. SRIM and GEANT4 allow estimating stopping power, particle range and energy straggling under a wide variety of experimental conditions.

References

[1] Ziegler J F 1999 Stopping of energetic light ions in elemental matter *J. Appl. Phys.* **85** 1249–72

[2] Barkas W H, Birnbaum W and Smith F M 1956 Mass-ratio method applied to the measurement of *L*-meson masses and the energy balance in pion decay *Phys. Rev.* **101** 778–95

[3] Bloch F 1933 Zur Bremsung rasch bewegter Teilchen beim Durchgang durch Materie *Ann. Phys.* **408** 285–320

[4] Berger M J and Seltzer S M 1964 *Tables of energy losses and ranges of electrons and positrons, NASA Special Publication* Order R-80, vol. NASA-SP-3012 National Bureau of Standards NASA

[5] Fano U 1963 Penetration of protons, alpha particles, and mesons *Annu. Rev. Nucl. Part. Sci.* **13** 1–66

[6] Ziegler J F, Ziegler M D and Biersack J P 2010 SRIM—the stopping and range of ions in matter (2010) *Nucl. Instrum. Methods Phys. Res.* B **268** 1818–23

[7] Bohr N 1940 Scattering and stopping of fission fragments *Phys. Rev.* **58** 654–5

[8] Firsov O B 1959 A qualitative interpretation of the mean electron excitation energy in atomic collisions *Sov. Phys. JETP* **9** 1076–80

[9] Jenkins D 2020 *Radiation Detection for Nuclear Physics: Methods and Industrial Applications* (IOP Publishing)

www.ingramcontent.com/pod-product-compliance
Ingram Content Group UK Ltd.
Pitfield, Milton Keynes, MK11 3LW, UK
UKHW051855100926
8556IPUK00005B/637